EDITED BY
ADI KUNTSMAN, SAM MARTIN
AND ESPERANZA MIYAKE

DIGITAL DISENGAGEMENT

COVID-19, Digital Justice and
the Politics of Refusal

BRISTOL
UNIVERSITY
PRESS

First published in Great Britain in 2023 by

Bristol University Press
University of Bristol
1-9 Old Park Hill
Bristol
BS2 8BB
UK
t: +44 (0)117 374 6645
e: bup-info@bristol.ac.uk

Details of international sales and distribution partners are available at
bristoluniversitypress.co.uk

British Library Cataloguing in Publication Data
A catalogue record for this book is available from the British Library

ISBN 978-1-5292-3465-7 hardcover
ISBN 978-1-5292-3466-4 ePub
ISBN 978-1-5292-3467-1 ePdf

The right of Adi Kuntsman, Sam Martin and Esperanza Miyake to be identified
as editors of this work has been asserted by them in accordance with the
Copyright, Designs and Patents Act 1988.

Cover design: Andrew Corbett
Front cover image: Panther Media GmbH / Alamy Stock Photo
Bristol University Press use environmentally responsible
print partners.
Printed and bound in Great Britain by CPI Group (UK) Ltd,
Croydon, CR0 4YY

FSC
www.fsc.org
MIX
Paper | Supporting
responsible forestry
FSC® C013604

Contents

Notes on Contributors

Chelsea Butkowski is Postdoctoral Fellow at the Center on Digital Culture and Society within the University of Pennsylvania's Annenberg School for Communication. Her research examines how people use digital media technologies to perform and represent their identities during periods of great social and technological change.

Adi Kuntsman is Reader in Digital Politics in the Department of History, Politics and Philosophy at Manchester Metropolitan University. Their research explores a range of topics including digital militarism, digital memory, digital refusal and environmental harms of digital communication.

Sam Martin is Senior Research Fellow at the Rapid Research and Evaluation Lab at University College London, as well as a senior digital analyst and digital sociologist at the Ethox Centre within the Big Data Institute at the University of Oxford. Her research focuses on the application of big qualitative data methodologies to analysing public health emergencies, as well as the impact of vaccine hesitancy and misinformation in social media settings.

Esperanza Miyake is Chancellor's Fellow in Journalism, Media and Communication at the University of Strathclyde. With a wide range of publications in media, cultural and digital studies, as well as related media and stakeholder engagements, Miyake specializes in examining issues relating to gender, race and technology.

Annika Richterich is Assistant Professor in Digital Culture in the Faculty of Arts and Social Sciences at Maastricht University. Her research explores social practices emerging in interaction with digital technology, and highlights the opportunities associated with hacker and maker spaces, and spaces/settings that minoritized groups have found(ed) for their engagement with digital technology.

Serra Sezgin is Assistant Professor and Head of the New Media and Communication Department at Ankara Science University. Her research and publications focus on new media, creative industries, labour and game studies.

Introduction

Adi Kuntsman, Sam Martin and Esperanza Miyake

In 2020, the world experienced an unprecedented global crisis that destabilized existing social, cultural, legal, economic, ethical, medical and digital structures in various contexts across different countries. The COVID-19 pandemic impacted lives in every society around the world. We now collectively face the task of critically assessing how these disruptions, destabilizations and decentralizations of power, justice and equality have created new opportunities, while also cementing and exacerbating existing inequalities. One of the key ways in which these opportunities and/or further oppressions were mechanized and operationalized was through the rapidly increased use of digital technologies across various spheres in science, technology and society: from public health, to education, politics, work and everyday life. In many instances, the newly introduced technologies were compulsory, or seen as necessary and unavoidable. Introduced as temporary measures, they often remained even after the initial 'need' had gone away. The rise in digitization was sudden and rushed, often justified by the crisis nature of the pandemic, leaving little room for critical interventions, reflection or resistance.

Within the context of these recent monumental and global changes brought by the pandemic, how then can we understand – and be critical of – new forms and practices of digital technologies and networked communications? How might they challenge, reinforce and/or maintain existing, normative inequalities, injustices and infringements to human and data rights around the world? What new dilemmas were created by shifting much of work, education and everyday interactions into online environments? What new data grabs have emerged or become normalized? And finally, what spaces and possibilities are there to resist encroaching digitality? What is the place of digital refusal in the fabric of pandemic and post-pandemic life?

Exploring these questions is what motivated us to put this book together, to begin making sense of the rapid and extensive increase, reliance and shifts in meaning of digital technologies after the pandemic. In order to critique some of the inequalities and consider the potential opportunities for digital justice, we situate our discussion within a larger theoretical framework of 'digital disengagement' (Kuntsman and Miyake, 2015; 2019; 2022). Developed as a paradigmatic shift that challenges the normalization of digital technologies in everyday life and social research, this framework begins at the point of disconnection, refusal and opting out. Assessing any form of digital transformation, innovation or governance from the point of view of whether it allows refusal, and to what extent, how and to whom, digital disengagement is a paradigm that puts digital justice at the heart of digital policy, practices and society. Having already worked on the rise of compulsory digitization and the shrinking spaces of opting out of digital health in the time before the pandemic (Kuntsman et al, 2019), we are driven by a sense of urgency, watching how the culture of 'technological solutionism' (Morozov, 2013) on the one hand, and the violence of digital coercion and discrimination (Gangadharan, 2017; Williams, 2018; Williams and Kind, 2019; Gangadharan, 2020) on the other, became normalized during the pandemic, embraced by many, and then cemented as necessity.

While acknowledging that many digital technologies and platforms were beneficial – if not crucial – for continuing social and economic operation, containing the virus, and supporting everyday life (especially during lockdowns), our primary focus here is to explore the less visible impacts of pandemic digitization. These include the increase in surveillance and infringements on data rights, shifts in working conditions and work–life balance, enforced digital participation, and a growing environmental footprint of digital infrastructures. The book's contributors address these through the lens of digital disengagement, by asking – or documenting – ways in which the negative impacts of pandemic digitization may be challenged, resisted and refused.

This book thus has two key aims. First, to explore digital disengagement and opt-out through the lens of the global pandemic, and second, to explore the pandemic and post-pandemic digital surge through the critical perspective of digital disengagement, disconnection and refusal. What are the implications of the further intensification of compulsory digitality since the pandemic? What are the relationships between increasing digitization and social injustice? What does digitization bring for future social, cultural and political formations in the post-pandemic world? What spaces will there be for digital disengagement, and what is the viability of opt-out from digital technologies, networks, tracing surveillance and databases? What forms of everyday and systemic resistance can we foster? What other, more just futures, can we imagine, through – or without – digital technologies?

Pandemic digitalities

Academic studies exploring the role that digital communication technologies played during the pandemic have already begun to emerge. Much of the work on digital connectivity has arisen, understandably, from the field of health, covering topics related to experiences of healthcare workers and patients, as well as broader debates on the medical and public health aspect of the pandemic; for example, an analysis of online narratives of adults with long COVID (Callard and Perego, 2021; Miyake and Martin, 2021), a discussion of the use of various types of social media by healthcare workers to share their mental health experiences as they worked on the pandemic frontline (Dowrick et al, 2021), and an analysis of online debates regarding vaccine and mask mandates (Martin and Vanderslott, 2022). At the same time, we are seeing a steady growth of scholarship in social sciences and humanities, exploring the social impact of the pandemic, particularly the importance of understanding global and local inequalities and their relationships to both digital technologies and the COVID-19 pandemic itself. For example, *The COVID-19 Crisis: Social Perspectives* (2021) by Deborah Lupton and Karen Willis captures the various global experiences of the pandemic, including the forms of marginalization involved. Similarly, a review by Rose-Redwood et al (2020) looks at the geographies of COVID-19, including issues surrounding digitalities and inequalities.

Most relevant to the concerns of this book are studies that explore pandemic digitization specifically with regard to digital surveillance and digital oppression. Most notable among these is Lupton's work on surveillance, containment and care (2021), Kaya's (2020) exploration of safety and privacy in relation to contact-tracing systems, Kitchin's (2020) discussion of civil liberties and pandemic surveillance, an investigation of a 'pandemic surveillance state' by Kampmark (2020), and Everts' (2020) notion of the 'dashboard pandemic'; as well as many other publications, many of which appeared in the first year of the pandemic as a rapid response to the unprecedently rapid changes taking place globally (Das, 2020; French and Monahan, 2020; Kouřil and Ferenčuhová, 2020; van Kolfschooten and de Ruijter, 2020; Yu, 2020). This book builds on these themes through the overall lens of digital disengagement, and the changing meaning of 'opting out' that the pandemic has instigated.

Finally, and perhaps most crucially, we should note two edited volumes that address the relationships between datafication, (in)justice and resistance: *COVID-19 from the Margins: Pandemic Invisibilities, Policies and Resistance in the Datafied Society* (Milan et al, 2021) and *Data Justice and Covid-19: Global Perspective*

(Taylor et al, 2020). Our book is inspired by, and offers an addition to these two hugely important volumes. We specifically look at questions of marginalization, oppression and inequality as a result of pandemic digitalities and coercion, not just as a result of the pandemic, but in themselves mechanisms of how the pandemic was digitalized at a global scale. We explore and conceptualize the tolls and dangers of digital acceleration and compulsory digitality through the perspective of digital refusal and disengagement.

The pandemic was indeed a global event changing the digital landscape across the world forever. While we cannot capture the global contexts – North and South, East and West – our book presents a series of geo-political contexts to provide a snapshot of how pandemic digitalities unfolded for citizens, users, activists and businesses in various ways. We see our book as an important intervention that centralizes digital disengagement, digital reduction and digital opt-out as a particular framework to address structural and everyday violences of the digital and as a way of thinking about resistance.

Digital disengagement

With the increasing proliferation of digital media and technologies centred around a logic of internet centricism and technological solutionism (Morosov, 2013), there has also been increasing interest in the concept and practice of digital refusal, disconnection and resistance to the digital. From data protection policies that support and safeguard opt-outs or digital detox retreats offered to those who feel overwhelmed by their digital lives, to platforms and services offering 'digital decluttering' and other forms of digital reduction, the idea of restricting or limiting digital technologies is clearly a growing agenda. The shift in public perceptions and practices are also reflected in the emerging academic field of 'Disconnection Studies', which explores the various implications, politics, economics and facilitations of disconnection from the digital.

Most research within Disconnection Studies focuses on user practices in online communication, with a particular emphasis on social media (Gershon, 2011; Karppi, 2011; Portwood-Stacer, 2012; Baumer et al, 2013; Portwood-Stacer, 2013; Karppi, 2014; Kaun and Schwarzenegger, 2014; Light, 2014; Light and Cassidy, 2014; Portwood-Stacer, 2014; John and Dvir-Gvirsman, 2015; Jorge, 2019).

Our concept of digital disengagement (Kuntsman and Miyake, 2015; 2019; 2022) is in direct dialogue with this body of work, but also critiques the tendency to naturalize the connection between 'digital' and 'social'. In other words, 'digital disengagement' moves beyond issues of 'social engagement' (and social media) and thinks through digitality in relation to other forms of forced engagement: it is about engagement with the political economy of digitality, its cultural formations, digital materiality, legal frameworks, and everyday actions that structure time and space. Digital disengagement thus theorizes disconnection itself as embedded within a Western-centred, capitalist and neo-liberal context of digitality that makes digital refusal and opt-out difficult for everyone (van Dijk, 2013; Hesselberth, 2018), while simultaneously othering those marginalized through digital architectures and everyday techno-practices. This book draws from our existing work on digital disengagement and actively pushes against the Western-centred approach to digital disengagement by considering how 'digital disconnection' is operationalized according to differences in culture, society, regional regulations and practices.

Furthermore, one of the key concerns of digital disengagement as discussed here is its emphasis on questions of digitality and social marginality, as related to compulsory digitality and 'digital coercion' (Gangadharan, 2020) and the 'encroachment of tech' (Williams, 2018) that primarily affects racialized, poor and marginalized communities. In that respect, one of the key themes in this book is the discussion of how the rise in digitization since the pandemic reproduces and intensifies existing forms of marginalization, injustice and systemic violence, placing digital justice at the heart of our enquiries. This issue is even more important now in the context of digitality, when 'digital coercion' has been packaged and presented as a collective necessity to ensure 'protection', 'safety' and 'containment of the virus'. Who is being protected by digital technologies, we ask, and who is being further marginalized, made more precarious or powerless? What kind of digital engagements did the pandemic create, who do they benefit and who do they harm? As we move towards a post-pandemic world, our question is twofold when we ask: what histories of inequality, injustice and violence feed current formations of compulsory digitization? What futures do pandemic digitalities bring? How might new digital systems and processes initially adopted as ad-hoc emergency responses to the pandemic be mishandled and mis-used; for example, digital population surveillance technologies adopted for the purposes of public health and virus tracing that are becoming further developed and normalized as technologies of everyday surveillance and policing. In that respect, the book explores not only the relationships between disconnection and social in/justice, but also specifically focuses on how structures, technologies and practices of injustice and marginalization are embedded, adopted, appropriated and mis-used within the context of the pandemic and post-pandemic life.

Chapter summaries

Looking at digital disengagement through the lens of COVID-19 and at the pandemic through the lens of digital justice, contributors to the book take a variety of perspectives: legal, social, cultural, political, technological and economic. All chapters examine the meaning of pandemic digitalities, the shrinking possibilities of refusal and opt-out, and new emerging tactics of digital disengagement. The book opens with a pair of chapters exploring the active and sometimes violent ways in which the question of digital engagement is forced upon individuals, communities and nations.

By focusing on some of the troubled ways in which the Tokyo 2020 Olympics unfolded amidst a global pandemic, Esperanza Miyake's chapter examines how digital engagement is racialized, reinforcing dominant discourses of power, race and digitality. The chapter begins by considering how the Tokyo 2020 Olympics represented a moment whereby digitality was hailed as a global and 'safe' solution to the problem of the pandemic, and yet how, behind this digital veneer promoting accessibility, equality and technological innovation, there were material and invisible consequences that were detrimental to questions of equality and human rights. Through an analysis of the relationship between media, the International Olympic Committee and Japan, the chapter offers the double notion of forcing and enforcing digitality: whereby digital engagement becomes a means of maintaining inequality through marginalization of the Other, while also propagating the capitalist-driven political global economy. Here, the discussion centres around the naturalization of digital, social and economic dis/engagement within a pandemic time–space that brought together the need for material labour, digital productivity and consumption. As such, this chapter demonstrates how digital engagement inevitably intersects with political, social and economic engagement, and, in the context of Tokyo 2020 Olympics, how it was ultimately weaponized, commodified and exploited by both the International Olympic Committee and Japan, as inter-related and mutually complicit forms of racialized digital solutionism, both serving their own 'project' of economic, political and socio-cultural dominance.

Serra Sezgin's chapter on the theme of enforcement focuses on forced digital engagements and work–life balance during the pandemic, adopting a self-reflective approach and autoethnographic method to understand the impacts and triggers of this forced digital engagement and its indications in daily ife. Sezgin kept diaries for two years (2019–2021) during which she held two different jobs involving creative labour, where she experienced different work models (working at the office, work from home and flexible working) that also allowed her to make comparisons. Her autoethnographic journey revealed that continuity of the need for know-how – which had an established position in the creative industries – made digital learning processes inevitable with the pandemic. While digital learning gains continuity with integration of each new digital tool into daily life; the obligatory self-taught, enforced adaptation to various digital platforms and networks by sharing data and spending time and effort involuntarily indicates that flexible capitalist exploitation is deepened by an expansion of unpaid digital labour. Consequently, especially for creative workers, the viability of opting out of digital technologies as a right should be defended against flexible capitalist exploitation disseminated in every aspect of our lives.

Sam Martin's chapter on digital disengagement and the repeal of *Roe vs Wade* frames the case of menstrual and fertility apps in the context of post-pandemic health app-ization and datafication, which led to increased digital surveillance and digital authoritarianism. Martin shows how acutely aware women are becoming with regard to how vulnerable apps that solicit personal data are making them in relation to state surveillance capitalism. The chapter uses social media discourse and sentiment analysis to understand how, in the immediate aftermath of the overturning of *Roe vs Wade*, women and their allies strove to effectively delete, digitally disengage, algorithmically disrupt or opt-out of sharing their data via period-tracking apps. In the context of an increased reliance on mobile health apps during the pandemic, analysis of this discourse on social media shows that women want to continue to use these apps to help them manage their period and fertility cycles; however, there is a strong desire to disengage from the threat of being prosecuted should they need an abortion in the process. Tracking practices of deletion, disruption and opt-out, the chapter theorizes digital disengagement as a form of resistance, addressing the ongoing tensions that arise between a more technologically informed public, with a right to opt-out and digitally disengage, and the need for governments to protect the public and their right to self-track their health without harsh, infringing and criminalizing regulation.

With an alternative, analogue perspective of digital disengagement, Chelsea Butkowski's chapter looks at how, amid the social distancing, doom-scrolling and mounting Zoom fatigue that accompanied the COVID-19 pandemic, people reached for pens, paper and postage stamps to write and send letters. Butkowski discusses how the practice of writing letters was shared on social media and how, the popularity of hashtags about the practice and aesthetics of letter writing paradoxically grew on the same connective platforms that handwritten letters circumvent. Overall, Butkowski found that pandemic letter writing and sharing contribute to the ongoing conversation on writing on- and off-line, by probing the boundaries of the on-off-line continuum. This, in turn, inspired questions about how technological apparatuses and interactional techniques come together to constitute a hybrid form of digital disengagement. Drawing together campaign materials as well as official Instagram accounts and Instagram posts, Butkowski adeptly investigates how people practise digitally disengaged pandemic communication by using the very digital technologies they are trying to challenge. Her chapter is an illustrative case for probing the definitional boundaries of digital disengagement as a continuum, rather than as a dichotomy of connection versus disconnection.

Continuing the topic of datafication and opting out, Annika Richterich's chapter on COVID-19 hacktivism explores a particular tactic of digital disengagement, which she terms 'data minimalism', in which excessive and expanding datafication is challenged for not being secure, safe, effective or efficient. Against the background of data expansion and 'dataism' that characterized the response to COVID-19 in Germany,

Richterich's chapter focuses on a hacktivist group that proposes systematic reduction of datafication through a range of practices that are informed and strategic, partial and selective – instead of merely refusing digital technologies entirely. Data minimalism and related hacktivism relate to longer histories of concerns about state surveillance and citizen freedom, and emphasize the importance of understanding both citizens' knowledge of digital surveillance and informed and deliberate micro-practices of citizen defence against the state.

Adi Kuntsman's chapter on digital solutionism similarly critiques the unjustified digital expansion and acceleration. However, unlike Richterich's focus on data expansion and privacy, Kuntsman's chapter primarily addresses the material and environmental tolls of infrastructure expansion that had a long pre-pandemic history but increased substantially from the start of the pandemic, while being rarely discussed or acknowledged. The chapter begins with a critique of the consistent and systemic blindness towards environmental footprints of digital technologies that characterized media and academic debates around pandemic digitization, despite the increased interest in climate change and ways that the pandemic may be affecting the environment. The chapter then explores the few academic interventions that did consider the environment, to promote a change in how we should understand the relationships between post-pandemic digitalities and social/environmental concerns. In doing so, the chapter points to the emerging language of 'digital sobriety', 'small file' art and digital reduction, and considers small but effective steps to reduce, rather than expand, the ongoing harms imposed by pandemic digitalities on both people and the environment.

Bringing together the critical questions discussed throughout the book, the volume concludes with a conversation between two leading scholars of digital technologies, datafication and social justice – Seeta Peña Gangadharan and Patrick Williams. In the conversation, curated by the book's editors, Gangadharan and Williams address key questions concerning state racism, socio-economic inequality, and over-policing of racialized and minority communities, and how these have intensified or shifted during the pandemic. They discuss how the encroachment of tech into state violence paradoxically goes hand in hand with the seductive power of tech to not only solve problems but create safety for the very marginalized communities who are subjected to violence, digital and otherwise. They consider how digital inclusion can be a form of state violence, and how community-based resistance needs to be based on digital self-defence. They explore the intensification of power, often shared by corporations and governments, but also consider the many possibilities of resistance and thinking ahead. Dissecting a simplistic idea of digital disengagement as resistance – for example, in contexts where refusal is a privilege and where challenging the violence of datafication and digital surveillance can lead to escalation and more violence – Gangadharan and Williams put forward the notion of digital abolition and what it takes to prepare for an abolitionist future in the present. This final chapter is followed by the epilogue in which we discuss the future of pandemic and post-pandemic digitalities.

References

Baumer, E.P.S., Adams, P., Khovanskaya, V.D., Liao, T.C., Smith, M.E., Schwanda Sosik, V. and Williams, K. (2013) 'Limiting, leaving, and (re)lapsing: an exploration of Facebook non-use practices and experiences', in E.P.S. Baumer, P. Adams, V.D. Khovanskaya, T.C. Liao, and M.E. Smith, et al (eds) *Proceedings of the SIGCHI Conference on Human Factors in Computing Systems*, Paris, France: ACM, pp 3257–3266.

Callard, F. and Perego, E. (2021) 'How and why patients made long covid', *Social Science & Medicine*, 268: 113426.

Das, S. (2020) *Surveillance in the Time of Coronavirus: the Case of Indian Contact Tracking App Aarogya Setu*. Datactive: The Politics of Data According to Civil Society.

Dowrick, A., Mitchinson, L., Hoernke, K., Mulcahy Symmons, S., Cooper, S., and Martin, S., et al (2021) 'Re-ordering connections: UK healthcare workers' experiences of emotion management during the COVID-19 pandemic', *Sociology of Health & Illness*, 43(9): 2156–2177.

Everts, J. (2020) 'The dashboard pandemic', *Dialogues in Human Geography*, 10(2): 260–264.

French, M. and Monahan, T. (2020) 'Dis-ease surveillance: how might surveillance studies address COVID-19?', *Surveillance & Society*, 18(1).

Gangadharan, S.P. (2017) 'The downside of digital inclusion: expectations and experiences of privacy and surveillance among marginal internet users', *New Media & Society*, 19(4): 597–615.

Gangadharan, S.P. (2020) 'Digital exclusion: a politics of refusal', in L. Bernholz, H. Landemore and R. Reich (eds) *Digital Technology and Democratic Theory*, Chicago: University of Chicago Press, pp 113–140.

Gershon, I. (2011) 'Un-friend my heart: Facebook, promiscuity, and heartbreak in a neoliberal age', *Anthropological Quarterly*, 84(4): 865–894.

Hesselberth, P. (2018) 'Discourses on disconnectivity and the right to disconnect', *New Media & Society*, 20(5): 1994–2010.

John, N.A. and Dvir-Gvirsman, S. (2015) '"I don't like you any more": Facebook unfriending by Israelis during the Israel–Gaza Conflict of 2014', *Journal of Communication*, 65(6): 953–974.

Jorge, A. (2019) 'Social media, interrupted: users recounting temporary disconnection on Instagram', *Social Media + Society*, 5(4): https://doi.org/10.1177/2056305119881691.

Kampmark, B. (2020) 'The pandemic surveillance state: an enduring legacy of COVID-19', *Journal of Global Faultlines*, 7(1): 59–70.

Karppi, T. (2011) 'Digital suicide and the biopolitics of leaving Facebook', *Transformations*, 20: 1–28.

Karppi, T. (2014) *Disconnect.Me: User Engagement and Facebook*, Turku, Finland: University of Turku.

Kaun, A. and Schwarzenegger, C. (2014) '"No media, less life?" Online disconnection in mediatized worlds', *First Monday*, 19(11): https://doi.org/10.5210/fm.v19i11.5497.

Kaya, E.M. (2020) *Safety and Privacy in the Time of Covid-19: Contact Tracing Applications*, Istanbul: Centre for Economics and Foreign Policy Studies.

Kitchin, R. (2020) 'Civil liberties *or* public health, or civil liberties *and* public health? Using surveillance technologies to tackle the spread of COVID-19', *Space and Polity*, 24(3): 362–381.

Kouřil, P. and Ferenčuhová, S. (2020) '"Smart" quarantine and "blanket" quarantine: the Czech response to the COVID-19 pandemic', Eurasian Geography and Economics, 61(4–5): 587–597.

Kuntsman, A. and Miyake, E. (2015) *Paradoxes of Digital Dis/Engagement: Final Report. 6. Working Papers of the Communities & Culture Network+*, Leeds, Sheffield and York: White Rose University Consortium.

Kuntsman, A. and Miyake, E. (2019) 'The paradox and continuum of digital disengagement: denaturalising digital sociality and technological connectivity', *Media, Culture & Society*, 41(6): 901–913.

Kuntsman, A. and Miyake, E. (2022) *Paradoxes of Digital Disengagement: In Search of the Opt-Out Button*, London: University of Westminster Press.

Kuntsman, A., Martin, S. and Miyake, E. (2019) 'Re-thinking digital health: data, appisation and the (im)possibility of "opting out"', *Digital Health*, 5. Available from: https://doi.org/10.1177/2055207619880671

Light, B. (2014) *Disconnecting with Social Networking Sites*, Basingstoke, UK: Palgrave Macmillan.

Light, B. and Cassidy, E. (2014) 'Strategies for the suspension and prevention of connection: rendering disconnection as socioeconomic lubricant with Facebook', *New Media & Society*, 16(7): 1169–1184.

Lupton, D. (2022) 'The Quantified Pandemic: Digitised surveillance, containment and care in response to the COVID-19 crisis', in S. Pink, M. Berg, D. Lupton, M. Ruckenstein (eds) *Experiencing and Anticipating Emerging Technologies*, London: Routledge, pp 59–72.

Lupton, D. and Willis, K. (eds) (2021) *The COVID-19 Crisis: Social Perspectives*, London: Routledge.

Martin, S. and Vanderslott, S. (2022) '"Any idea how fast 'It's just a mask!' can turn into 'It's just a vaccine!'"': from mask mandates to vaccine mandates during the COVID-19 pandemic', *Vaccine*, 40(51): 7488–7499.

Milan, S., Treré, E. and Masiero, S. (eds) (2021) *COVID-19 from the Margins: Pandemic Invisibilities, Policies and Resistance in the Datafied Society*, Amsterdam: Institute of Network Cultures.

Miyake, E. and Martin, S. (2021) 'Long Covid: online patient narratives, public health communication and vaccine hesitancy', *Digital Health*. 7. Available from: https://doi.org/10.1177/205520 76211059649

Morozov, E. (2013) *To Save Everything, Click Here: Technology, Solutionism and the Urge to Fix Problems That Don't Exist*, London: Allen Lane.

Portwood-Stacer, L. (2012) 'How we talk about media refusal, part I', Flow [online]. Available from: http://www.flowjour nal.org/2012/07/how-we-talk-about-media-refusal-part-1/ [Accessed 3 August 2018].

Portwood-Stacer, L. (2013) 'Media refusal and conspicuous non-consumption: the performative and political dimensions of Facebook abstention', *New Media & Society*, 15(7): 1041–1057.

Portwood-Stacer, L. (2014) 'Care work and the stakes of social media refusal', Critical Personas [online]. Available from: http://www. newcriticals.com/care-work-and-the-stakes-of-social-media-refu sal/prin [Accessed 1 December 2019].

Rose-Redwood, R., Kitchin, R., Apostolopoulou, E., Rickards, L., Blackman, T., and Crampton, J., et al (2020) 'Geographies of the COVID-19 pandemic', *Dialogues in Human Geography*, 10(2): 97–106.

Taylor, L., Martin, A., Sharma, G. and Jameson, S. (2020) *Data Justice and Covid-19: Global Perspective*, Manchester: Meatspace Press.

van Dijk, J. (2013) *The Culture of Connectivity: A Critical History of Social Media*, Oxford: Oxford University Press.

van Kolfschooten, H. and de Ruijter, A. (2020) 'COVID-19 and privacy in the European Union: a legal perspective on contact tracing', *Contemporary Security Policy*, 41(3): 478–491.

Williams, P. (2018) 'Being Matrixed: the (over)policing of gang suspects in London', Stop Watch [online]. Available from: https://www.stop-watch.org/our-work/gangs-matrix [Accessed 12 March 2019].

Williams, P. and Kind, E. (2019) 'Data-driven policing: the hardwiring of discriminatory policing practices across Europe', European Network Against Racism [online]. Available from: https://www.enar-eu.org/Reports-Toolkits-153 [Accessed 4 September 2020].

Yu, A. (2020) 'Digital surveillance in post-coronavirus China: A feminist view on the price we pay', *Gender, Work & Organization*, 27(5): 774–777.

ONE

En/forcing the Tokyo 2020 Olympics: The Racialization of Digital Engagement and Digital Solutionism

Esperanza Miyake

Introduction

On 24 March 2020, the International Olympic Committee (IOC) announced that both its president, Thomas Bach, and the Japanese Prime Minister at that time, Shinzo Abe, had reached an agreement to postpone the Tokyo 2020 Olympics by a year, in response to the COVID-19 pandemic crisis. The statement presented the decision as one made between the IOC and Japan, for the 'good' of the people, stating that 'Human lives take precedence' and 'We agreed that the safety, security of the athletes and spectators are paramount' (International Olympic Committee, 2020). A year later, with COVID-19 cases rising in Japan (which had to declare a state of emergency weeks before the Olympics), despite calls from the medical community and the majority of the population to cancel, the Olympics nonetheless went ahead. Athletes competed in audience-free stadias, and the promised boom for the tourism industry and the Japanese economy never occurred. Furthermore, there was little to no option for

Japan to refuse: Japan was legally and contractually obliged to make the Olympics happen, or else they faced great financial loss/heavy fines. Global broadcasters, corporate sponsors and spectators around the world, watching from the comfort of their screens, had everything to gain. Japan was still burdened with the reality of having to deliver the Olympics for global consumption, despite the great associated financial losses and risks to the health and wellbeing of its citizens in the middle of a global pandemic.

The only way in which the Tokyo 2020 Olympics was able to happen at all was largely due to Japan's digital infrastructure. The Tokyo 2020 Olympics represents some of the most alarming – and as we will see, problematically racialized – ways in which the digital was offered as a perfect 'solution' to the 'problem' of hosting a global event during the pandemic. In his work on technological solutionism, Morosov (2013) discusses the propensity for solutionists to recast complex social situations through quick technological 'fixes', arguing that such 'solutions' lead to 'unexpected consequences that could eventually cause more damage than the problems they seek to address' (Morosov, 2013: p 5). In the case of Tokyo 2020, taking place right in the middle of a global pandemic, the idea of 'damage' is one that gains even greater significance, as such damage relates not just to the economy but to human life, rights and welfare. The digitalization of Tokyo 2020 might have 'saved' the Olympics, might have 'solved' the problem of a potential cancellation ,but the question is who benefited from this form of digital solutionism (Kuntsman and Miyake, 2019; 2022). Was it those safely at home being entertained through an internet connection? Was it the broadcasting corporations who made millions from broadcasting rights? Was it Bach and the IOC who ensured the continuation of a sporting legacy and whose own authority remained intact through promised delivery?

Furthermore, what about the material consequences? In Tokyo, $15.4 billion (if not more) was invested into providing a digital experience globally (Pesek, 2021), yet local and national businesses suffered financially from the lack of tourism (for example, Japan Airlines lost $2.6 billion during the pandemic, see Shibata, 2021; The Japan Times, 2021). Similarly, the journalists and athletes on the ground had challenging experiences physically and emotionally navigating the event (Taku and Arai, 2020; Majumbar, 2021), and COVID-19 cases increased due to the influx of Olympics-related travellers and the struggle to contain the virus at such a large-scale event, despite the measures and protocols in place (Cox, 2021; McCurry, 2021a). Such material consequences seem to have been overlooked and conveniently blamed on COVID-19 as the culprit, rather than the enforced digital solutionism as the direct cause. After all, it is not COVID-19 that enforced the Tokyo 2020 Olympics: it is digital solutionism that became a material means of reinforcing social, economic and political inequalities.

This chapter starts from this disjunction between the digital and the material, arguing how this tension points toward the racialization of digital engagement in ways that reinforce and maintain dominant ideologies of power, race and digitality. In the process, I explore the double notion of both forcing and enforcing digitality – one that moves away from the idea of the neoliberal agent responsible for their own digital dis/engagement – whereby digital engagement becomes a means of maintaining inequality through the marginalization of the Other, stabilizing and naturalizing the uneven power structures facilitated by digitality. I am particularly interested in unpicking how such processes of digital engagement intersect with issues relating to social and economic engagement, especially within the context of the pandemic where such connections have been delineated through shifting practices in everyday life (such as social distancing and the online economy boom). Ultimately, I argue that the Tokyo Olympics 2020 represents a form of racialized digital engagement and digital solutionism, where digitality is both weaponized and commodified as a means to consolidate political, socio-cultural and economic power, hidden under the convenient guise of pandemic safety and wellbeing.

Pandemic digitalities: digital, social and economic engagement

There is a growing body of work – increasingly referred to as 'Disconnection Studies' – that considers the complex tensions within individual and/or collective relationships with digitality (Portwood-Stacer, 2013; Karppi, 2014; Kaun and Schwarzenegger, 2014; Light, 2014; Light and Cassidy, 2014; Hesselberth, 2018; Jorge, 2019), ones that issues relating to agency, identity, surveillance, justice and/or digital economies and labour (van Dijck, 2013; Fuchs, 2014, 2015; Zuboff, 2019). In my past work (Kuntsman and Miyake, 2022), I critiqued the naturalization of the social with the digital. Increasingly, the very idea of the 'digital' (and, by extension, 'the internet') seems to have become entangled with the 'social', largely because of the proliferation of social media across all areas of life: 'Western' society is increasingly moving towards being 'social media ready', part of an overall process of platformization (Helmond, 2015; van Dijck et al, 2018). As such, this social mediatization – an intersectional sibling of globalization, mediatization abd platformization – has increasingly become the sole means of understanding 'social engagement', always measured through the metricization of social relationships through quantitative analyses (number of followers, likes, re-posts, replies and so on). Furthermore, because the digital economy increasingly relies on the conflation of social media, social engagement and digitality – think of the gig economy, targeted advertising, data mining – the idea of 'digital engagement' is not only about social engagement but also about engagement with wider structures of digital capitalism: hence, the digital becoming an increasingly compulsory form of governmentality that regulates sociality, economy, citizenship and identity. We are thus now at a point where we do not quite know how to conceptualize and/or practice social engagement without the digital, and, similarly, profit without digital/social engagement.

Within this context, Tokyo 2020 presents us with a case that complicates such machinations, simultaneously reinforcing but also destabilizing the naturalized link between digital, social and economic engagement. Hailed and promoted by the IOC as one of the most innovative and globally connected Olympic Games, Tokyo 2020 is a prime example whereby the digital presented the financial and practical solution – masquerading as global engagement and connectivity – to the pesky little problem of holding a mass event during a global, life-endangering pandemic, with the IOC issuing statements such as 'Fans to enjoy innovative digital experiences for Tokyo 2020' (International Olympic Committee, 2021a) and 'Unprecedented broadcast coverage and digital innovation to connect fans around the world to the magic of Tokyo 2020' (International Olympic Committee, 2021b). Tech and Information and Communications Technology (ICT) stakeholders were equally enamoured by the digital productivity of Tokyo 2020, where technology 'saved the day', with articles such as 'Tokyo Olympics: the most connected Games ever' (Poll, 2021), 'Tokyo 2020 Olympics inspires hi-tech advancements' (GlobalData, 2021), and 'Game changer: the first Olympic games in the cloud' (Zhang, 2021). Such media discourse highlighting 'connectivity' reveals how the digital has further embedded itself into the concept and practice of social engagement.

However, paradoxically and ironically, Tokyo 2020 was a largely people-less affair where athletes competed in audience-free stadias, and where what little in-person social engagement there was, was highly regulated in accordance with COVID-19 restrictions. Unlike social media and online engagement which is expected to be virtual, the Olympic Games – like most sporting events – have traditionally been associated with emotive, 'live', in-person audiences cheering, where the quality of the experience is based on the material co-presence, co-participation and co-production of emotion, identity and spectatorship, to a point where the digital is even perceived as a hindrance to this material engagement (Hutchins, 2016). What is interesting is that Tokyo 2020 was a moment where popular media discourses acknowledged the absence of this crucial material/social engagement, but somehow presented the digital (solution) as compensating for this through digital–social/economic engagement in terms of viewing figures and social media innovation (it was the first Olympics to hold a TikTok account, for example): typical comments include 'Tokyo 2020 played out in empty arenas but bound eight billion people together' (Bull, 2021). In other words, in the eyes of media and tech corporations (and of course the IOC), Tokyo 2020 was an example of digital triumph over pandemic adversity: a digital solutionist approach to social and economic engagement, despite the material dangers of the quickly spreading COVID-19. This kind of pandemic capitalism exacerbated inequalities through uneven distribution of safety and labour across the material and digital divide, while presenting economic profit to those who could keep the pandemic digitally at bay:

> In contrast to the safer-from–home digital workers who inhabit the virtual and largely asynchronous temporalities of the web, 'essentiality' is bound to the 'real time' of deliveries, services and frontline work that take place in the synchronous and inhabited geographies with close proximity to infection. The pandemic exacerbates the ongoing crisis of capitalism by parcelling out who must suffer in real time and who lives in virtual time. (Chan, 2020: p 13.5)

The stubborn blindness to and disconnect between the digital and material reality of Tokyo 2020 is deeply concerning, and one that is founded upon financial and political gain over human life and the welfare of a nation. In discussing the inequalities played out through the shifting configurations of time, space, digitality and labour during the pandemic,

Chan (2020) discusses the idea of 'distal temporalities', where productivity, opportunity and capitalist time reinforce inequalities in the techno-caste system (Chan, 2020; p 13.2). Indeed, while the digital and virtual space was seen as a contactless, non-material and thus safe solution to the problem of a usually materially saturated global sporting event, the question of 'safe' depends on one's geopolitical orientation: safe for whom? Is it safe for the 'safer-from-home' spectators and the decision-makers consuming the digital product? Or is it safe for those having to risk their lives physically in Tokyo to produce the digital product for mass consumption? Tokyo 2020 thus marks a moment at which material productivity and digital consumption were situated on unequal terms amidst a pandemic 'techno-caste' system based on profit and politics.

But what is perhaps the most concerning issue here is the forcing and enforcing of digital engagement. On the whole, Disconnection Studies explores largely Western individual and/ or collective techno-practices of resistance against the digital that are conscious in some way; whether by leaving social media, going on digital detoxes, engaging in hacktivism/algorithmic jamming, most forms of digital disengagement and refusal come from a privileged position rooted within neoliberal capitalist structures that provide the concept and reality of agency, the exercising of 'choice' and (self)responsibility. My previous work (Kuntsman and Miyake, 2022) critiqued this configuration, highlighting the unequal distribution of temporal, digital, spatial and economic capital, arguing ultimately how the elastic continuum of digital disengagement meant that, while some may enjoy digital detoxes and phone-free holidays, for others the spaces and opportunities for opt-out were fast shrinking.

However, Tokyo 2020 presents us with something even more sinister: here, it is not just the shrinking space of opt-out as previously postulated, but also an active penalty for opting out, imposed through a legal and financial obligation that does not take into account the extraordinary pandemic circumstances. Even during one of the worst global pandemics in recent history, Tokyo 2020 was not allowed to be cancelled, and, indirectly, was forced to commit to the solution of digital engagement, enforced by the threat of legal fines that would be imposed by the IOC and related Olympic broadcasters and sponsors (McCurry, 2021b). The digital solution offered and the penalties of opting out put Tokyo 2020 in checkmate. Here, digital disengagement becomes not just a 'choice'/non-option similar to the situation of platform labourers who rely on the gig economy and for whom digital disengagement is not really a choice as it leads to financial loss and/or increased risk to wellbeing/health, but also a non-option with the threat of an active, material, disciplinary punishment that is internationally visible. There is no choice, or even a pretense of 'choice', only the active punishment of choice.

The question of financially imposing a fine for opting out is incredibly alarming as it moves beyond questions of self-discipline and a lack of agency and into violent impositions of forced labour, risk to life and exacerbations of inequality: this is not compulsory digitality, it is forced and enforced digitality. Tokyo 2020 thus represents the weaponization of the digital as both a solutionist facilitator for economic and political gains by those in power and a truncheon of civil, political, legal and financial obedience despite the high risk to human life. There is an increasing body of work that explores weaponization of the digital and digital technologies as a way of gaining political, social and economic power while oppressing the Other (O'Neil, 2017; Noble, 2018; Benjamin, 2019; Eubanks, 2019). Tokyo 2020 marks a moment where the oppression lies not so much in the digital weapon itself, but in its forced use by profit-driven corporations and the IOC as an international institution: a globally enforced digitality and productivity, to serve all those who are power-, money- and entertainment-hungry. Digitality was thus an exchangeable political, economic and cultural currency in the unequal international transaction between global politics, global sports and global media.

The racialization of digital engagement and digital solutionism

In the previous section, I critiqued the ways in which the digital 'solution' of producing Tokyo 2020 was one that was not only forced and enforced, but also one that was grounded upon geopolitical, disjunctive inequalities of engagement and disengagement. I discussed the configuration and naturalization of digital, social and economic dis/engagement as situated within a pandemic time–space that brought together the need for material labour, digital productivity and consumption. In this final section, I want to take a closer look at how such a process was simultaneously enabled by and reinforced existing digital (self)orientalist ideologies (Miyake, 2022) surrounding Japan as an East Asian country. How can we understand digital solutionism in the context of (self)Orientalist and capitalist ideologies? How is digital engagement and digital solutionism racialized?

Morley and Robins (1995) wrote their seminal piece on techno-Orientalism about a time when Japan represented a threat ('the Japan panic') through its technological dominance over the West. For the authors, the simultaneous exoticization and fear of the Other may be understood through the West's relationship with Japan and its technological innovation, production and dominance, in which technology represents an ideological, economic and socio-cultural destabilization and decentralization of Western power. Referring to Said's (1978) theories on Orientalism, the authors argue that 'The 'Orient' exists because the West needs it; because it brings the project of the West into focus' (Morley and Robins, 1995: p 155). The technology of the Other' represents a potential threat to this project. Thinking through such issues in the context of technological supremacy, as being simultaneously economic, cultural and political, for Morley and Robins (1995), the concept of techno-Orientalism is thus formed at the ideological crux of techno-myth and the economic/political materiality of Japanese technologies representing futurity and innovation on the one hand, while also representing the West's technological emasculation leading to its cultural, political and economic defeat.

The realisation of Morley and Robins' concept of 'techno-Orientalism' relies on the materiality of technology, the material conditions that enable production of its ideological practices that give raise to racism and technological raciality. For example, there would be no 'Japan-bashing' without the material technologies embodying socio-cultural values that led to the practice of physically bashing material commodities. However, since the 1990s, and certainly in the era of Tokyo 2020, there has been a significant shift from techno-centricism to digi-centricism. This shift has in many ways made invisible the inequalities that arise from the materialities involved in the production of digitality (Qiu, 2016); as discussed earlier, Tokyo 2020 is a perfect example where digitalization was hailed as innovation and the solution during the global pandemic, but rendered invisible the risks and dangers associated with the local materialities of digital production.

Discussing platform and racial capitalism in the context of digital infrastructures, economies and societies, McMillan Cottom (2020) explores the idea of predatory inclusion, defined as 'the logic, organization, and technique of including marginalized consumer-citizens into ostensibly democratizing mobility schemes on extractive terms (McMillan Cottom, 2020; p 443)'. As we have seen from the official IOC and news media discourse presented earlier, the Tokyo Olympics were rigorously promoted as using digital technologies and platforms as a means to open up unprecedented 'accessibility' and 'inclusion' for all. Here, digitality (and engagement with it) is aligned with democracy, equality and racial parity. Conveniently, there is already a wealth of recent digital–Orientalist discourses that naturalize the relationship between 'the digital', 'smart' and 'AI' as something Japan can 'do' so easily (Miyake, 2022). Japan does it again! Japan is 'Westernized'!

However, ultimately, such co-optation was part of a racialized political economy that benefitted only those countries that had the digital infrastructures, rights and streaming services to broadcast the Olympics, and the media and broadcasting corporations themselves. In fact, even within the Global North/the 'West', where digital infrastructures are most advanced, digital engagement and thus access was not a given, and had to be bought, a classed process that further naturalizes the relationship between digital engagement and financial engagement (and thus, indirectly, social engagement). For example, in the UK, the BBC did not have the full rights to broadcast the Olympics, having been outbid by US broadcasting company Discovery, who won the European rights and placed most events behind a paywall (Waterson, 2021). Similarly, in Japan itself, public broadcaster NHK won the rights to televise Tokyo 2020, but this was as a joint venture with six other private broadcasters (Nippon Television, TV Asahi, Tokyo Broadcasting System (TBS) Television, Fuji Television, TV Tokyo, and J:COM TV) who made content available only to subscribers (Chandra, 2021). In both cases, where is the 'equality' and 'accessibility' in digital engagement?

Similarly, as well as the financial benefits gained by mega corporations such as Discovery, forcing the Tokyo Olympics to go ahead at all literal and figurative costs was also ideologically empowering the IOC, as an 'organisation that has been instrumental in garnering support for a form of international cooperation clothed in the resuscitation of the Olympic idea' (Chappelet and Kübler-Mabbott, 2008; p xiii). In this context, the IOC thus represents a project that capitalizes on and promotes global engagement (through economic, political and digital engagement) while obsfucating racial inequality. Digital engagement provides the perfect screen to hide the racialized economy of difference, while also providing a front that promises political, cultural and social engagement.

Indeed, weeks before the eventual Olympics unfolded in July 2021, Bach referred to Japanese citizens as 'Chinese people'.

Within the context of the COVID-19 pandemic, which provoked violent attacks against East and Southeast Asians worldwide that led to significant East and Southeast Asian anti-racist activism (Phillips, 2020) such a 'gaffe' (a term many news media outlets used) is extraordinary. Yet, as for many politicians and people in power before him, the news story died down, and Bach remained unscathed, ultimately waving flags, handing out medals and bouquets during the ensuing Olympic ceremonies in Tokyo. Despite the Olympics and the IOC being supposedly apolitical, neutral and standing for 'global values' (as exemplified by Rule 50, which imposes a ban and heavy consequences for any athletes displaying any allegiance towards a political movement), such moments reveal the asymmetries in power that are underscored by contradiction – we oppose racism but ban any reference challenging it; we will punish athletes standing against racism but will not punish our own members who display racist behaviour – ultimately marginalizing and disadvantaging the Other. Is this the 'Olympic ideal'?

It is tempting of course to simply define the IOC as the dominant 'Master of Ceremonies' and Japan as the submissive host, whereby the obvious alignment of racial difference and inequality can be assigned to the IOC/Bach (White) and Japan/Abe (Other), in their respective binary oppositionalities. However, to take a Foucauldian approach here, power is not something that is exclusively held and exercised by the IOC. Japan's (or perhaps originally, Abe's) own position and decision to even engage with the IOC and Olympic socio-politics and economics was part of its broader strategic plan to reform Japan's constitution (Klein, 2020). We must not forget that it was also originally Abe's determination to strengthen Japan's position within the global political and financial landscape – part of his 'Abeconomics' policies – that led to Tokyo becoming the Olympics host city: 'showing off' the nation's digital and smart capabilities (Waldenberger, 2020) was a central part of this plan, inevitably linking the digital with political, socio-cultural and economic global engagement. The Tokyo government estimated the Games would give a ¥32 trillion boost to the national economy (Lewis and Murad, 2020), something that would work in conjunction with its strategic self-exploitation of soft power: who can forget Abe dressed up as Super Mario at the Rio 2016 Olympics?

Furthermore, although by the time the Olympics eventually occurred, Abe had stepped down as Prime Minister for health reasons (Yoshihide Suga took his place in 2020), Abe was still very vocal leading up to the event in his championing of the Olympics as the right course of action for Japan, despite mass protests across the country. Discussing the anti-Olympics movement in Japan, Ganseforth (2020) explores how the Olympics in many ways brought together those who were marginalized and/or otherwise fighting for their rights: from the exploitation of workers on the Olympic construction sites to homeless people and those living in public housing being dislocated from the Olympic vicinity as a result of privatization of public spaces. The anti-Olympics movement was also concerned about increased securitization and surveillance – marking further restrictions of civil liberties – under the pretext of deploying anti-terrorist strategies at the Olympics (Ganseforth, 2020). This is a classic case whereby digital engagement is paraded as the safe solution for 'the good of society', while operationalizing infrastructures that not only exacerbate inequalities but insultingly rely on the oppression of the marginalized. Although there was media coverage of such national dissent, on the whole, there was what has been described as the Japanese media's 'manufacturing of national consent' right from the early stages of Japan's bid in 2016 (Manzenreiter, 2020: p 10).

As a retort to anti-Olympics sentiment arising in Japan in 2021, Abe indirectly suggested that those opposing the Olympics are, in effect, opposing Japan and its own financial/cultural project for internationalization: 'People who have been criticized by some as anti-Japan because of their historical perceptions and other views are now strongly opposing staging the Olympics' (Tadokoro, 2021). Although this may not necessarily be considered 'racist', as one might Bach's Chinese/Japanese 'gaffe', such statements nonetheless point towards the racialization of engagement. In this case, Abe's views around the supposed anti-Japanese sentiment circulating within Japan emerge from his own nationalistic project: a complicit engagement with political, economic and socio-cultural power as facilitated by the digital. Japan's digitality is used as a tool to prove that it has the advanced means to 'save' and deliver the Olympics, while also as a commodity to showcase its position as a world leader in technological innovation. Digital engagement for Japan and the Olympics was thus part of its own self-Orientalizing discourse (Miller, 1982; Iwabuchi, 1994; Moeran, 1996; Miyake, 2022), in which naturalization of 'digital' and 'Japan' was encouraged, while serving the practical purpose of materializing the Olympics 'safely' during the pandemic.

In this context, digital engagement thus becomes a mechanism of governmentality that enforces Japanese discipline, obedience and subservience, not just towards the IOC or the Olympics but to Japan as a nation. Within such a configuration, opting out of digitality suddenly becomes not just a politics of resistance against questions such as mass surveillance and digital rights, but also becomes embroiled with questions of national identity, if not (anti)nationalism. In turn, such a process becomes part of a process of a self-racialization of Japanese-ness through digital engagement and digital solutionism: both serve as a means and product to materialize and self-perpetuate the myth of 'Japan', ultimately requiring the othering and silencing ('manufacturing of consent') of those material Japanese Others who oppose digital engagement.

Conclusion

By focusing on the Tokyo 2020 Olympics, this chapter has explored the ways in which the racialization of digital engagement and digital solutionism reinforces and maintains dominant discourses of power, race and digitality. Exploring the simultaneous divorcing and obfuscating of materialities from the digitalization involved in realising the Olympics, I first explored the concept of enforcing digitality – one that moves away from the idea of the neoliberal agent responsible for their own digital dis/engagement – whereby digital engagement becomes a means of maintaining inequality through marginalization of the Other, whilst also propagating the capitalist-driven political global economy. I critiqued the forcing and enforcing of the Tokyo 2020 Olympics was grounded upon geopolitical, disjunctive inequalities of engagement and disengagement, and discussed the configuration and naturalization of digital, social and economic dis/engagement as situated within a pandemic time–space that brought together the need for material labour, digital productivity and consumption. As such, I argued that digital engagement inevitably intersects with political, social and economic engagement, and how this was ultimately weaponized, commodified and exploited by both the IOC and Japan, as inter-related and mutually complicit forms of racialized digital solutionism, both serving their own 'project' of economic, political and socio-cultural dominance.

It seems fitting to end this chapter by acknowledging the shocking assassination of Shinzo Abe in 2022. Bach and the IOC of course led tributes, lowering the Olympic flag to half-mast at Olympic House in Lausanne for three days following the news (International Olympic Committee, 2022). Words like 'partnership' and 'friendship' are used in the official statements released by the IOC at the time. Upon first glance, such words may be dismissed as mere official rhetoric. However, they also highlights the fact that the racialization of digital engagement and digital solutionism of the Tokyo 2020 Olympics should be contextualized within Orientalist ideologies: part of both the IOC/the 'West's project to embody and incorporate the Other through its technologies and digitalities, a form of digi-Orientalism, and part of Japan's self-Orientalist digital strategies and ideologies that intersect at the politics and economics of national identity and nationalism. In this sense, the inequality rests not so much between Japan and the IOC, but between Japan/IOC and the silent Others who were either co-opted and digitally (en)forced to engage, or did not have the means to engage politically, economically or socially despite their obedient digital engagement.

References

Benjamin, R. (2019) *Race After Technology: Abolitionist Tools for the New Jim Code*, Cambridge, UK: Polity Press.

Bull, A. (2021) 'Tokyo 2020 played out in empty arenas but bound eight billion people together', *The Guardian* [online] 8 August. Available from: https://www.theguardian.com/sport/2021/aug/08/tokyo-2020-played-out-in-empty-arenas-but-bound-eight-billion-people-together [Accessed 16 February 2022].

Chan, N. (2020) 'Pandemic temporalities: distal futurity in the digital Capitalocene', *Journal of Environmental Media*, 1(Suppl): 13.1–13.8.

Chandra, C. (2021) 'How to watch the Tokyo 2020 Olympics from home', *Tokyo Weekender* [online] 21 July. Available from: https://www.tokyoweekender.com/2021/07/how-to-watch-the-tokyo-2020-olympics [Accessed 10 July 2022].

Chappelet, J.L. and Kübler-Mabbott, B. (2008) *International Olympic Committee and the Olympic System (IOC): The Governance of World Sport*, London: Routledge.

Cox, C.M. (2021) 'Counting cases, counting medals: containing the Olympic contagion during the Tokyo Games', in D. Jackson, A. Bernstein, M. Butterworth, Y. Cho, D. Sarver Coombs, M. Devlin and C. Onwumechili (eds) *Olympic and Paralympic Analysis 2020: Mega Events, Media, and the Politics of Sport. Early Reflections from Leading Academics*, Brussels: European Communication Research and Education Association, p 29.

Eubanks, V. (2019) *Automating Inequality: How High-Tech Tools Profile, Police, and Punish the Poor*, New York: Picador.

Fuchs, C. (2014) *Digital Labour and Karl Marx*, London: Routledge.

Fuchs, C. (2015) *Culture and Economy in the Age of Social Media*, London: Routledge.

Ganseforth, S. (2020) 'The difference between zero and one: Voices from the Tokyo anti-Olympic movements', in B. Holthus, I. Gagné, W. Manzenreiter and F. Waldenberger (eds) *Japan Through the Lens of the Tokyo Olympics*, London: Routledge, pp 110–115.

GlobalData (2021) 'Tokyo 2020 Olympics inspires hi-tech advancements', GlobalData [online] 13 August. Available from: https://www.globaldata.com/tokyo-2020-olympics-inspires-hi-tech-advancements-says-globaldata/ [Accessed 16 February 2022].

Helmond, A. (2015) 'The platformization of the web: making web data platform ready', *Social Media + Society*, 1(2). Available from: https://doi.org/10.1177/2056305115603080.

Hesselberth, P. (2018) 'Discourses on disconnectivity and the right to disconnect', *New Media & Society*, 20(5): 1994–2010.

Hutchins, B. (2016) "We don't need no stinking smartphones!' Live stadium sports events, mediatization, and the non-use of mobile media', *Media, Culture & Society*, 38(3): 420–436.

International Olympic Committee (2020) 'Tokyo 2020 Olympic and Paralympic Games postponed to 2021', International Olympic Committee [online] 24 March. Available from: https://olympics.com/en/news/tokyo-olympic-games-postponed-ioc [Accessed 11 February 2022].

International Olympic Committee (2021a) 'Fans to enjoy innovative digital experiences for Tokyo 2020', International Olympic Committee [online] 21 July. Available from: https://olympics.com/ioc/news/fans-to-enjoy-innovative-digital-experiences-for-tokyo-2020 [Accessed 16 February 2022].

International Olympic Committee (2021b) 'Unprecedented broadcast coverage and digital innovation to connect fans around the world to the magic of Tokyo 2020', International Olympic Committee [online] 9 July. Available from: https://olympics.com/ioc/news/unprecedented-broadcast-coverage-and-digital-innovation-to-connect-fans-around-the-world-to-the-magic-of-tokyo-2020 [Accessed 16 February 2022].

International Olympic Committee (2022) 'IOC mourns the death of former Japanese Prime Minister Abe Shinzo', International Olympic Committee [online] 8 July. Available from: https://olympics.com/ioc/news/ioc-mourns-the-death-of-former-japanese-prime-minister-abe-shinzo [Accessed 30 July 2022].

Iwabuchi, K. (1994) 'Complicit exoticism: Japan and its other', *Journal of Media & Cultural Studies*, 8(2): 49–82.

The Japan Times (2021) 'JAL reports first net loss since 2012 relisting as pandemic bites' [online] 7 May. Available from: https://www.japantimes.co.jp/news/2021/05/07/business/corporate-business/jal-business-year-losses/ [Accessed 15 February 2022].

Jorge, A. (2019) 'Social media, interrupted: users recounting temporary disconnection on Instagram', *Social Media + Society*, 5(4). Available from: https://doi.org/10.1177/2056305119881691.

Karppi, T. (2014) *Disconnect.Me: User Engagement and Facebook*, Turku, Finland: University of Turku.

Kaun, A. and Schwarzenegger, C. (2014) '"No media, less life?" Online disconnection in mediatized worlds', *First Monday*, 19(11). Available from: https://doi.org/10.5210/fm.v19i11.5497.

Klein, A. (2020) 'Political games', in B. Holthus, I. Gagné, W. Manzenreiter and F. Waldenberger (eds) *Japan Through the Lens of the Tokyo Olympics*, London: Routledge, pp 13–17.

Kuntsman, A. and Miyake, E. (2019) 'The paradox and continuum of digital disengagement: enaturalizing digital sociality and technological connectivity', *Media, Culture & Society*, 41(6): 901–913.

Kuntsman, A. and Miyake, E. (2022) *Paradoxes of Digital Disengagement: In Search of the Opt-out button*, London: Westminster University Press.

Lewis, L. and Murad, A. (2020) 'Japan: how coronavirus crushed Abe's Olympics dream', Financial Times [online] 30 March. Available from: https://www.ft.com/content/c343aa5e-702a-11ea-9bca-bf503995cd6f [Accessed 15 February 2022].

Light, B. (2014) *Disconnecting with Social Networking Sites*, Basingstoke, UK: Palgrave Macmillan.

Light. B. and Cassidy, E. (2014) 'Strategies for the suspension and prevention of connection: rendering disconnection as socioeconomic lubricant with Facebook', *New Media & Society*, 16(7): 1169–1184.

Majumdar, B. (2021) 'How Covid changed sport – a case study of the 2020 Tokyo Olympic Games', *Sport in Society*, 26(1): 184–190.

Manzenreiter, W. (2020) 'Olympics and the media', in B. Holthus, I. Gagné, W. Manzenreiter and F. Waldenberger (eds) *Japan Through the Lens of the Tokyo Olympics*, London: Routledge, pp 8–11.

McCurry, J. (2021a) 'Pride and anxiety mingle in Japan as Olympic medal tally and Covid cases rise', The Observer [online] 8 August. Available from: https://www.theguardian.com/sport/2021/aug/08/pride-and-anxiety-mingle-in-japan-as-olympic-medal-tally-and-covid-cases-rise [Accessed 15 February 2022].

McCurry, J. (2021b) "Claims could run into billions': the interests at stake if Olympics in Japan were cancelled', *The Guardian* [online] 10 June. Available from: https://www.theguardian.com/sport/2021/jun/10/claims-could-run-into-billions-the-interests-at-stake-if-olympics-in-japan-were-cancelled [Accessed 10 July 2022].

McMillan Cottom, T. (2020) 'where platform capitalism and racial capitalism meet: the sociology of race and racism in the digital society', *Sociology of Race and Ethnicity*, 6(4): 441–449.

Miyake, E (2022) 'I'm a Virtual Girl from Tokyo: Virtual Influencers, digital-Orientalism and the (Im)materiality of Race and Gender', *Journal of Consumer Culture*, 23(1): 209–228.

Miller, R.A. (1982) *Japan's Modern Myth: The Language and Beyond*, New York: Weather Hill.

Moeran, B. (1996) 'The Orient strikes back: advertising and imagining Japan', *Theory, Culture & Society*, 13(3): 77–112.

Morley, D. and Robins, K. (1995) *Spaces of Identity: Global Media, Electronic Landscapes and Cultural Boundaries*, London: Routledge.

Morosov, E. (2013) *To Save Everything, Click Here: the Folly of Technological Solutionism*, New York: Public Affairs.

Noble, S.U. (2018) *Algorithms of Oppression: How Search Engines Reinforce Racism*, New York: New York University Press.

O'Neil, C. (2017) *Weapons of Math Destruction: How Big Data Increases Inequality and Threatens Democracy*, London: Penguin Books.

Pesek, W. (2021) 'Tokyo Olympics shows why Japan wins gold for bad economy', Forbes [online] 27 July. Available from: https://www.forbes.com/sites/williampesek/2021/07/27/tokyo-olympics-shows-why-japan-wins-gold-for-bad-economy/?sh=2bf4d3274efe [Accessed 15 February 2022].

Phillips, A. (2020) "Coronavirus given the face of an East Asian person': Parliament debates racism during pandemic', *Sky News* [online] 13 October. Available from: https://news.sky.com/story/coronavirus-given-the-face-of-an-east-asian-person-parliament-debates-racism-during-pandemic-12102897 [Accessed 20 February 2022].

Poll, S. (2021) 'Tokyo Olympics: the most connected games ever', International Telecommunication Union [online] 10 August. Available from: https://www.itu.int/hub/2021/08/tokyo-olympics-the-most-connected-games-ever/ [Accessed 16 February 2022].

Portwood-Stacer, L. (2013) 'Media refusal and conspicuous non-consumption: the performative and political dimensions of Facebook abstention', *New Media & Society*, 15(7): 1041–1057.

Qiu, J.L. (2016) *Goodbye iSlave: A Manifesto for Digital Abolition*, Champaign, IL: University of Illinois Press.

Said, E.W. (1978) *Orientalism*, London and Henley: Routledge & Kegan Paul Ltd.

Shibata, N. (2021) 'Amid pandemic, JAL posts biggest loss of $2.6bn since relisting', Nikkei Asia [online] 7 May. Available from: https://asia.nikkei.com/Business/Transportation/Amid-pandemic-JAL-posts-biggest-loss-of-2.6bn-since-relisting [Accessed 12 May 2022].

Tadokoro, R. (2021) 'Ex-PM Abe says 'anti-Japan' people are strongly opposing Tokyo Olympics', *The Mainichi* [online] 3 July. Available from: https://mainichi.jp/english/articles/20210703/p2a/00m/0na/009000c [Accessed 10 June 2022].

Taku, K. and Arai, H. (2020) 'Impact of COVID-19 on athletes and coaches, and their values in Japan: repercussions of postponing the Tokyo 2020 Olympic and Paralympic Games', *Journal of Loss and Trauma*, 25(8): 623–630.

van Dijck, J. (2013) *The Culture of Connectivity: A Critical History of Social Media*, Oxford: Oxford University Press.

van Dijck, J., Poell, T. and de Waal, M. (2018) *The Platform Society*, Oxford: Oxford Academic.

Waldenberger, F. (2020) 'The Olympic and Paralympic Games as a technology showcase', in B. Holthus, I. Gagné, W. Manzenreiter and F. Waldenberger (eds) *Japan Through the Lens of the Tokyo Olympics*, London: Routledge, p 136.

Waterson, J. (2021) 'BBC Olympics coverage misses events after loss of TV rights', The Guardian [online] 25 July. Available from: https://www.theguardian.com/media/2021/jul/25/bbc-olympics-coverage-misses-events-after-selling-tv-rights? [Accessed 10 June 2022].

Zhang, J. (2021) 'Game changer: the first Olympic games in the cloud', MIT Technology Review [online] 8 September. Available from: https://www.technologyreview.com/2021/09/08/1035178/game-changer-the-first-olympic-games-in-the-cloud/ [Accessed 16 February 2022].

Zuboff, S. (2019) *The Age of Surveillance Capitalism: The Fight for the Future at the New Frontier of Power*, London: Profile Books.

TWO

Digital Engagements and Work–Life Balance in Creative Labour

Serra Sezgin

Introduction

The rapid change and transmission of communication technologies have transformed our interaction types and engagement styles on digital media parallel to social, political, cultural and economic relationships. As there is a dialectic relationship between cultural, social and daily life and digital transformation, the concept of digital engagement as a complex phenomenon requires consistent re-evaluation/re-thinking as for concepts such as digital economics, labour politics and inequality issues (Kuntsman and Miyake, 2019). In addition, as a result of the COVID-19 pandemic starting in 2020, the general population had to adapt to a new way of life, primarily because of the reduction of face-to-face communication, socialization and other physical interactions. The pandemic dramatically affected the social life and work practices of the general population. This game-changing situation created a need for more comprehensive and multidimensional approaches to communication studies, specifically the practices of digital (dis)engagement. The pandemic had worldwide socio-cultural impacts, but these new adaptations to social life and work culture also depended on national policies. Nations' COVID-19 regulations are mainly driven by their economic and technological capacities.

When the first COVID-19 case in Turkey was announced on 11 March 2020, the public and private sectors had already shifted to a flexible working system. Businesses with a small number of personnel in the workplace provided flexibility to their staff through remote working options. By 11 April, Turkey had announced a total lockdown for its citizens. Some measures, including the curfew, were taken by the government at various intervals, considering some factors such as seasonal conditions, holidays, age group, location, etc.

A year after the first COVID-19 case, Turkey was in 5th position in the world in terms of the number of cases that had been seen, and in 19th position in terms of mortality caused by COVID-19. During this period, the government tried to minimize the adverse economic effects of COVID-19 by providing low-interest loans or various support packages to its citizens. However, this period has been considered the biggest economic crisis era after World War II by the president in terms of it its outcome (Timeturk, 2020). In this period, when life was housebound, creative industries that used digital technologies at the centre and did not necessarily rely on a workspace or the physical presence at the office of the employees, brought workspaces to homes.[1] Remote work, home-office or hybrid working methods applied experimentally created new challenges such as maintaining work–life and/or work–leisure balance for workers. From this point of view, as a social scientist and a creative worker, I discuss in this chapter how the digital engagement experiences of creative workers have been transformed by the pandemic process and its effect onwork–life balance.

Creative industries and labour

The idea of creativity gained value in the 1980s with neoliberal ideas and policies of capitalism that embraced information technologies and entrepreneurship. Neoliberal ideology is compatible with capitalism's individualism, insecurity and risk-taking ideas, and its ambiguity is seen as a disciplinary factor guiding economic actors (Haiven, 2016). The idea of creativity came to be seen as a part of public discourse and the production process in the 1990s. This gave rise to a new industrial term, 'creative industry' that was directly associated with human creativity (Lee, 2017).

However, it is unclear under which conditions labour is considered 'creative labour', as there is no consensus regarding which industries are defined as creative industries or the key characteristics that determine creative industries. If we consider creativity as a feature inherent in human beings, everything that people do includes creativity and thus all kinds of labour are creative. However, in this study, to emphasize the conceptual and theoretical transformation of creativity and to encourage creative workers who hesitate to define themselves as workers and indirectly position themselves in capitalist exploitation and to be self-reflective about what this transformation brings, I critically use the terms creative labour and creative industry. I define creative labour as:

> a form of labor [in which] self-actualization is carried out with the motivation of love and passion, employees rush to work mostly voluntarily by identifying with the job, relatively autonomous, individual, flexible and precarious work is seen, and depending on these, work spreads to every moment and space of life and exploitation is legitimized. (Sezgin, 2020: pp 107–108)

Accordingly, four distinct components of creative labour that integrate and overlap may be summarized as (1) motivation, (2) time and space, (3) the production process, and (4) exploitation.

The link between creative labour and digital engagement in the pandemic may be more clearly defined by these components, which will be described in detail in forthcoming sections. To begin with, creative workers, whose main motivation is self-actualization and self-development (rather than financial gain), internalize work that is mostly flexible and autonomous and involves increased intensification of individualization and the situation of solitude, which hardens to reveal the relationships of exploitation and a collective action or solidarity against this (Sezgin, 2020). In this sense, as a determining factor in identifying the self and a way of life, what is defined as creative labour has different reflections in the context of work–life balance. Since the start of the COVID-19 pandemic, working from home, especially among creative workers, has increased the blurring of the work–life balance, and deepened the precarity. The legitimization of this precarity that lies in the so-called creative nature of labour now seems to be shifted to digital availability, or simply having an internet connection. Opting out of digital networks was considered as an option for at least a certain time before the pandemic; for example, to be able to socialize for a weekend or a night out, or when there is no signal available. As we stayed at home during the pandemic, we were expected to remain always connected, online and engaged, and to keep working and socializing through digital platforms. At some level, digital engagement is required to interact with the rest of the world. However, in the pandemic, it became more of a compulsion than a necessity or voluntary action. The change in the level and forms of digital engagement under pandemic conditions also relates to the work–life balance and the transformation of daily life, especially in creative labour in which the distinctions between leisure and work are already blurred. Therefore, in this study, I aim to understand the transformation in daily life, the work–life balance of creative labour and employees, through the 'new' forms that digital engagement took during the pandemic.

Methodology

In this research, I took a self-reflective approach and used an autoethnographic method to understand the effects of digital engagement and its indicators in daily life triggered by COVID-19, focusing on work–life balance. As a social scientist, I find it valuable to try to understand society and the individual in the context of communication, culture and political economy, in line with my epistemological and methodological approach, and to shed light on social inequalities (and reproduction practices of social inequalities). Therefore, autoethnography, which is a methodology in which the researcher places deep personal and emotional experiences in the context of broader social phenomena, serves this study well by interweaving the personal, the biographical, the political and the social (Denzin, 1997). Ellis and Bochner (2000: p 739) theorized autoethnography as the study of texts in which 'concrete action, dialogue, emotion, embodiment, spirituality, and self-consciousness are featured, appearing as relational and institutional stories affected by history, social structure, and culture, which themselves are dialectically revealed through action, feeling, thought, and language'.

In the present study, autoethography offered me the chance to create a dialogue (between myself, the text and the readers) and a common ground, while constructing my personal story against the background of the theory and the literature.

This study has a limited national context, mostly due to the methodological approach and scope. In any case, experiences of crises (such as the pandemic) should not be universalized or assumed to lead to the same self-reflections or struggles with digital (dis)engagements and work–life balance for everyone (Bagger and Lomborg, 2021). However, I believe autoethnography offers a connection beyond national dimensions, a common ground that involves consciousness of class and social change. Autoethnography is seen by Stacy Holman Jones as a method of making the personal political: 'setting a scene, telling a story, weaving intricate connections among life and art, experience and theory, evocation and explanation ... and then letting go, hoping for readers who will bring the same careful attention to your words in the context of their own lives' (2005: p 765).

As Denzin points out regarding the aim of autoethnography (2003: p 203), the basis of this work will be 'showing rather than telling' how forced digital engagement eliminated the acts of disengagement or opting out of digital devices as options for me by taking a reflexive look at my own narrative.

Eight months before the pandemic, I started to keep autoethnographic diaries covering two jobs over a period of two years. In 2019, after six years of working as a research and teaching assistant in various universities, and the day after completing my PhD, I was again working in the same field as my doctoral research but not as a researcher, rather as an employee in the advertising industry, a popular pillar of the creative industries. In 2020, as the pandemic continued, I returned to the university as an assistant professor. Thus, the research timeline includes a pandemic, two jobs involving creative work[2], and various working models such as the office, working from home and flexible working, allowing me to make comparisons and connections between digital engagement and my work–life balance. In the following sections, I will try to convey the subject by quoting from my diaries, and then, as Holman Jones explained, I hope to lay the groundwork for self-reflexivity for readers who I believe will be easily empathetic.

Digital engagements and work–life balance

The components of creative labour intersect with the daily life experience of (forced) digital engagement on many levels, in many forms. This makes it difficult to frame a fully synchronized narrative. Thus, the study is framed focusing on the moments, angles and contexts in which creative labour and digital engagement are intertwined. In this section, precarious conditions, alienation, individualization, flexibility and aspirational labour are discussed as the distinguishing elements of creative work in the context of digital (dis)engagements (including the viability of opting out of digital technologies as a right). In the following sub-sections, I will analyse the exploitative elements of creative labour that are successfully hidden under the discourse of self-development, productivity, and the empowerment of digital engagements.

Self-motivated: fast and injurious

I entered the advertising industry, where the turnover rate is extremely high, at the age of 30 (which is above the average), with a doctorate but without a portfolio. After I was accepted to the job from the entry position in the social media team, I had to prove my 'fast-learning' skill as soon as possible, and I had to adapt the work pace set in the academy to the rhythm of my current sector. Thanks to the collaborative approaches of my teammates, I soon discovered that physical endurance and speed are the main challenges, while accessing new information and learning are not very challenging. My job represented the invasion, consumption, dependency and coerciveness of 'offline' daily life by 'online' life due to the timelessness of digital engagement, which requires handling time in line with the digital flow. Due to the instantaneous nature of the digital, and its relationship with speed and flow, insomnia and body aches such as headache, eye pain and backache caused by long working hours in front of the screen and in the field became my primary problems: 'I will explain in very short keywords. New job: I'm exhausted. I'm dying. Chaos. I got mixed up. Pain. I do not want to see any phone or screen. I have to go to bed before 10 pm because I have to wake up early.'

I developed practical solutions to these problems within a short time. I hired a dog walker for my dog's morning walks. I bought a hair styler that dries and combs my hair at the same time. I stopped eating breakfast. I shortened the time on the road by sharing with a friend who has a car. In short, I created more time to work by minimizing the time spent on myself. When I look back on that period, during which I was proud of my solution-oriented approach, the first thing I see was that all these 'practical' solutions I produced increased my monthly expenses. I used to spend time making money and used money to save time.

Learning a second profession caused me to ignore one of the distinguishing components of creative work: motivation. Creative labour refers to a work process whereby the individual achieves their desire through the will of self-actualization and creates a new professional identity, which turns into a form of existence with a dedication to work. Creative labour does not always correspond to wage labour. In creative labour, the individual willingly adds dimensions to the product or production process by establishing an emotional connection with the work at various levels and in various contexts. Depending on the emotional and affective relationship in question, the sources of motivation and gain are feelings such as a passion for the work, love, pleasure and self-actualization.

Although my physical adaptation got stronger over time, the dramatic increase in my digital engagement rate started to reflect on my life in various dimensions, as I noted in my diary:

> My work is not finished on time. It's like I'm not alive. Recently, I felt the need to look at the sky intensely, like hunger, like thirst, a need to look at something endless, because I felt out of breath. I hope this feeling is temporary. I guess I can get used to it, I can speed up so I will have time to live in the future.

I was indeed speeding up, but there was something else I overlooked; as digital advertising, which is a fundamental gear of the creative industries, requires continuous learning and staying up-to-date, there was no end to the acceleration (Ashton, 2011). As long as the pace of the digital did not slow down, the pace of working could not be slowed down, and I had to quickly access the most up-to-date information and innovations and continue to improve myself.

As I aimed to move from social media to digital marketing, digital learning (Seymour et al, 2020), which includes the development of qualities such as reflexivity, agility and adaptability, and complex requirements, was a necessity, not only to fulfil my current job description but also for my career development in the sector. In other words, in addition to social media platforms such as Instagram, Facebook, Twitter and YouTube, digital learning platforms were also included in my digital engagement.

Exploitation as another distinctive component of creative labour has emerged parallel with the growth of 'heavy', tiring', 'difficult' and 'sleepless' processes, which are inevitable as time and continuity of self-development take on an invasive structure. The fact that the primary motivation of the employees is not financial gain makes it difficult to draw attention to exploitative relationships in the creative labour process. In creative labour, exploitation takes a flexible, implicit and voluntary form due to the precarious and flexible nature of the work and the dedication of the creative worker (Sezgin, 2020).

Engaged to all: the time and space

In line with capitalist production and neoliberal principles, precarious and flexible work is not just normalized day by day but even seems attractive. At this point, 'time and space' appear as another creative labour component. With the flexibility of work in the context of time and space, workplaces and working hours are often where/when workers also socialize or entertain. The distinction between work and leisure becomes blurred as working time spills over into every moment of life. In short, the time and space of creative labour are often not limited to a specific workplace and working hours. This is reflected in my experience mostly through the problem of 'know-how', which has an essential role in the development of creative industries. In the context of information, access and time, it also became my main problem, not just a problem for the industry. While developing the practice of co-production with a focus on solidarity, sharing knowledge and experience are the only factors that facilitate access to know-how and self-learning (Sezgin, 2020). However, when we switched to working from home due to the pandemic in March 2020, the reorganized working culture eliminated the possibilities of solidarity and interaction by making 'all in one' employees[3] both all in one and alone.

Due to the pandemic, the process of working from home limited interaction with colleagues and industry players, and therefore the opportunities to learn from each other and to learn together, while making engagement with 'know-how', digital platforms and networks more necessary than ever. It was no longer a choice to become a member of various networks, blogs, and platforms in order to gain know-how, rather it was essential to interact with colleagues, increase productivity, and make remote working effective. Online meeting tools, one of the most striking examples of digital engagement in the process of working from home, have been the tools that organized both informal and formal communication in work culture throughout the pandemic. This is quite normal considering that engagement could only happen digitally during the pandemic. On the other hand, because the practice of working from home worries employers in terms of productivity, online meetings often become unnecessary and/or long, increasing the workload and becoming a control and surveillance mechanism. In other words, surveillance and control of the productivity and efficiency of employees started to take place through online meetings or in-house digital communication channels. The use of many tools for file sharing, employee engagement, online meetings, task management, instant messaging, data storage, employee directory software, and many more have transformed the digital engagement of all employees. Since the use is mandatory, includes mandatory data sharing and voluntarily digital labour, with these increasingly diversified and multiplied digital tools that require separate digital learning processes, life was becoming more digital.[4]

In addition to digital platforms such as Instagram and WhatsApp, platforms for personal and private use were also turning into tools for work-related goals. Based on my own experience, when the working day starts (8 am), tens of employees sent a 'good morning' message on Slack one after the other to convey the message 'I woke up, I'm online and I'm working', or communicated our mobile phone numbers to students at the university on the pretext of protecting effectivity in online distance education, or shared or announced meetings via a WhatsApp group instead of emailing. Use of these applications, which is often based on justifications such as facilitating and speeding up work, increasing productivity or strengthening communication, undoubtedly results in the expectation of digital availability and accessibility of the employee at all times. Thus, the viability of opting out of digital networks is becoming more and more impossible.

Accordingly, we can say that the naturalization of engagement as digital (Kuntsman and Miyake, 2019), legitimized during the pandemic process, poses a challenge to the possibility of digital disengagement as a right or choice. At this point, when digital engagement has lost its voluntariness or preferability and taken on a forced appearance, I think it is extremely controversial whether all these applications increase productivity and efficiency. Similarly, as the knowledge that you are constantly at home and online due to the pandemic means that you do not have the necessary excuse with the manager and/or customers for not being able to complete a task, constant availability of information through digital media and the feelings that such availability creates are an additional problem (Kuntsman, 2012). From my point of view, looking forward to the hours when I can exercise my right to digital disengagement, hours when I don't feel the responsibility of constantly looking at the screen and being triggered, and when being asleep is considered normal, is an example of the emotions created by constant availability.

I remember that I wished to get sick enough to be unconscious for a week or something so that I could leave the digital devices/digital sphere with a solid excuse. With the necessary reflexivity today, I can see that digital disengagement, even for a while, meant unconsciousness or illness to me. I perceived engagement as living itself; for me, digital became the default (Kuntsman and Miyake, 2019). Any type of technology non-use as a means of empowerment or resistance, or digital disengagement on any level, seemed unlikely to me (Hesselberth, 2018). Just as I produced practical solutions to speed up my mornings before the pandemic, I stopped being interested in my Instagram blog, but this time without planning. Seeking a purification from my digital engagement during the pandemic, I started unfriending people I didn't know or had difficulty remembering for the sense of cleansing that it provided (Gershon, 2011; John and Dvir-Gvirsman, 2015).

Individualization and alienation: flexible and autonomous

The working models (remote/rotational) that were applied in the pandemic suggest flexible and autonomous work that is adapted to the components of creative labour and that has already been legitimized in creative industries. Although autonomous work may at first glance seem like freedom, it actually involves more authority and responsibility, neoliberal forms of governmentality such as self-management and self-discipline. While flexible and autonomous working practices in line with efficiency and productivity goals pave the way for self-control and intense individualization, they can turn into a control and surveillance mechanism (Sennett, 2013). In addition to their main workload, reorganizing the work constantly and synchronizing their private life with the organization becomes another obligation of the creative worker (Sezgin, 2020).

I also experienced working autonomously in my social media job due to working from home, and in my university job as a result of scientific production practices.[5] During the transition period, I wrote in my diary: 'I have to make new plans, but I'm still tired'. Making new plans and developing new strategies, which depend on the self-management mechanism, emerged as jobs in themselves. The reason why I turned this into an individual responsibility was that I personally took responsibility for any disruption or the slightest decrease in my productivity. Therefore, I was ready to produce surplus value in every sector, institution and under all conditions, as a creative worker who was involved in a self-directed, disciplined, supervised and controlled process and fully embraced the neoliberal subjectivity desired by capitalism. As I wrote in my diary: 'The workplace is pushing me hard. I can't progress in the project as I want. I have a blockage. I must throw it away'. Thus, there was no longer any need for anyone else to assign me tasks, monitor my workload, or expect me to increase productivity because I was doing all this to myself in the most intense and cruel way. While it wasn't clear who was making a profit in this situation, it was clear who was being exploited. My work–life balance has become completely invisible as a result. Responsibility for continuous self-improvement, self-control and self-management had created significant stress in my life in addition to my current workload. Instead of the average 8 hour shift, I was working more than 12 hours a day when the time I spent on self-learning and self-development is included. And all my time was intertwined with digital devices: online and connected.

Individualization is also realized at the point of flexible working, whereby creative workers undertake self-management in terms of motivating and disciplining themselves to control their own labour process (Sezgin, 2020). Bauman (2017) states that, in this sense, flexibility infiltrates every area of individual life (p 202) and in a life governed by the principle of flexibility, life strategies and future plans have to be short-term, temporary and variable (pp 204-205). Therefore, business life, which has always been full of uncertainties, appears today as a 'violent individualization movement' (Bauman, 2017: p 218). Progress is no longer a collective problem but an individual problem. Individuals are expected to use their own wisdom, resources and skills to leave their grievances behind and elevate themselves to more satisfying conditions (Bauman, 2017). In this sense, fears, anxieties and pains that do not come together to form a 'common cause' are experienced alone. This limits the possibility of solidarity (Bauman, 2017). As I wrote in my diary:

I am working so hard so that I can work less in the future. Little by little I began to think if I wanted this. Coronavirus, quarantine, illness ... Am I afraid to face the fact that there is nothing that defines me except work, that I have nothing to do but work? In fact, it was obvious that this would happen. Not voluntarily, but somewhat consciously, I let myself go. In a way, I gave up.

I studied creative labour and flexploitation (Bourdieu, 2017) and self-exploitation (Hesmondhalgh and Baker, 2011) during my PhD. I have done the research and written hundreds of pages about why we need to reproduce work–life balance, ourselves, subjectivity in a labour process other than work, and save our lives from the invasion of work. Yet, as you can see, I gave up. The alienation created by the intense individualization that I experienced came to the fore at this very point.

Given that, the 'production process', including relatively autonomous production, creative thinking (such as innovation, problem solving, and artistic skill), and individualization in various contexts, is revealed as the last distinctive component of creative labour. At this level, individualization is at its highest level by associating the success or failure of work with the patience, ambition, internalization of risk and individual struggle of the employee. The bond of creative labour with collective production is shaken by the solitude of the worker at this point (Sezgin, 2020).

Pursuing the balance: productivity strikes back

The components of creative labour related to neoliberal principles such as individualism, flexibility and self-actualization affect the level and forms of digital engagement and thus the work–life balance. The high level of digital engagement during the pandemic is used by the capitalist reproduction of neoliberal governmentality not just in terms of productivity at work but also in terms of leisure, by interfering with private life on a personal level. In this context, I describe my digital engagement experience on two basic lines connected to social and personal needs during the pandemic. Initially, I observed a boost in the discourse of staying healthy/fit/motivated on social media, which I believe creates social pressure in everyday life. Every time I saw digital content mentioning tips to increase productivity while working from home, things to do so as not to get bored, how to stay healthy and fit at home, I was thinking about the impossibility of finding time to 'get bored' at home. Never mind making bread at home or exercising, I couldn't find time to go to the toilet, which is five steps from my study. In a short time, my enthusiasm for content for those who were bored at home turned into anger that I directed first at those who produced such content and then at myself.[6]

On the one hand, we can easily criticize the reproduction of ideal capitalist subjectivities mediated by digital media through self-management, self-control, self-improvement and body image, even under global pandemic conditions. On the other hand, being exposed to this discourse through an intense bombardment of digital content 24/7 under the conditions of that period caused me to feel guilty for not being able to manage my time properly, not doing yoga, and not eating healthily despite everything, because I couldn't find time for anything other than work. It was as if I was the only one who could not make time for these activities, so I felt I must be doing something wrong.

C. Wright Mills (1951: pp xvi) writes of white-collar people that, when they get hired, they sell not only their time and energy but also their personality: 'For security's sake, he must strain to attach himself somewhere, but no communities or organizations seem to be thoroughly his.' It was in the mid-20th century when Mills argued that this isolated position made white-collar workers vulnerable to being shaped by popular culture, long before digital engagement at current levels was conceived. When we think about it from the perspective of digital engagement, the self-alienation that accompanies the alienation of labour also illuminated my own thoughts:

> Like many white-collar workers in the world, I drink alone at home. What drives us all to this loneliness? I used to think that I was living life to the fullest, but it's like I'm not living right now. It's like my life is shifting. It's like it's not my life, like my life is suspended, my days seem to be wasted. It's like I'm wasting my time.

Not being able to find a moment or space in my life to articulate myself confirms Mills' argument. The lifestyles promoted on social media during the pandemic process deepened this feeling. Not being able to position myself in the digital flow that had taken over my life made it even more impossible for me to root myself and find both space and time. It was a search for 'the non-fluid', 'the inflexible', 'the constant'. It was obvious that I was alienated from my labour process, my own life, and myself. What is worse is that I perceived it as 'wasting time'. While the flow of time and space disrupted the work–life balance, not being able to catch the course of this mobility and speed destroyed my positioning in life (Virilio, 1998). Even the discomfort of this situation was due to the fact that it hindered productivity.

The anxiety about productivity was also manifested in my second engagement line indirect from work during the pandemic: online dating. I resort to online dating apps as a way to escape self-alienation, intense individualization and loneliness. At this point, although digital engagement was inevitable in terms of both work and life, it is remarkable that escape (from the digital) also included the digital. In other words, seeking social engagement and digital disengagement at the same time was not possible. The digital and the social collapse into a singular, interchangeable concept in such a way that digital connectivity and engagement are defined and naturalized through the concept of social practice (Kuntsman and Miyake, 2019). Consequently, the possibility of opting out of digital and digital disengagement, constraining social relationships, was exhausted as well as my chance to escape.

Aside from that, it should be emphasized that productivity also took over online dating, which was my last line of defence in digital engagement. A little romance, a little socializing through online dating, which I took to for the purpose of sharing, was my only leisure time activity. According to Baudrillard's definition, the only factor that determines the freedom to lose time, kill it if necessary, spend it as pure waste (p 181), is that it is different from working time, defined by the absence of working time (Baudrillard, 2012: p 187). Therefore, online dating was more than a romantic pursuit, it was an effort to save my life from the invasion of work.

Indeed, André Gorz, in the *Critique of Economic Reason*, emphasizes that leisure activities have a rationality opposite to economic activities: 'Leisure is a time of pleasurable activity, which does not serve any purpose other than itself, where instrumental rationality categories such as efficiency, productivity, and success cannot be applied, and which only seduces' (Gorz, 1995: p 18). Whereas (my) 'economic reason' has instrumentalized the last branch to which I cling: when my online dating adventure did not reach its primary goal, I decided to conduct scholarly research on online dating because I couldn't let my time go to waste. In the absence of a romantic relationship, it would at least have served an academic purpose. Ultimately, my self-alienation was so deep that I did not allow myself the luxury of wasting any moment of life, I could not find free time for myself, or I did not consider myself worthy of it. All my life activities had turned into productive activities like those Erin Duffy (2016) has described as 'aspirational labour', with the potential to pay off in terms of future economic and professional opportunities.[7]

Conclusion

In this study, I have tried to rethink online technologies and digital engagement, which make work an invasive, compelling and consuming factor (Gregg, 2011), with the support of autoethnographic methods and concepts that reveal the voluntary affective dimensions of labour such as creative, aspirational and digital labour.

First, I must state that this first autoethnographic study was more difficult than I had anticipated. While a person rarely likes their reflection in the mirror, looking at themself through the eyes of an investigator, questioning, listening and keeping a distance, can be a deeper journey than we imagine.[8] The pandemic conditions may have taken each of us on similar journeys towards ourselves, in a way. This study may be read as an autoethnographic journey in which I, as a creative worker, try to understand how the pandemic affected my work–life balance, taking into account the components of creative labour under these conditions that directly affected the level and forms of digital engagement.

In this journey, I observed that the continuity of the need for know-how, which has an established position in the creative industries, made digital learning processes inevitable during the pandemic. While digital learning gains continuity with integration of each new digital tool into daily life, the obligatory self-taught, enforced adaptation to various digital platforms and networks by sharing data, spending time and effort involuntarily indicates that flexible capitalist exploitation is deepened by an expansion of unpaid digital labour (Fuchs, 2015).

During the pandemic, it seemed that digital engagement becam the default, with an invasion of leisure and private digital platforms by public and work-related ones. When the timelessness of digital flow and the spacelessness of the pandemic come together, the distinction that allows a balance between work and life disappears from the start. Based on their research conducted in Denmark with knowledge workers in the early pandemic, Bagger and Lomborg (2021: p 175) argue that, although it has advantages, working from home 'involves missing out on a lot of scaffolding and structuring of work tasks and this resulted in people experiencing problematic losses of boundaries'.

The forced appearance of digital engagement emerges as digital engagement motivation becomes work-centreed, digital platforms shift to work-related functions, and flow and speed in digital engagement are in a position to determine daily life. In my case, as my digital disengagement attempts covered the life-centred aspects, achieving a work–life balance became more and more unlikely. At this point, I should ask why these dynamics sremained when I changed my job and went back to work from the office. I suppose working from home during the pandemic is not the same when there is no life outside the home or office. Thus, what home represents has changed as a result of work having invaded the privacy that is usually expected in the home (Bagger and Lomborg, 2021). That was the first element that ruined the work–life balance. The second is related to anxiety about lack of productivity.

The surveillance and control mechanisms developed by employers because of their anxiety about lack of productivity reluctantly started to be adopted by the creative workers themselves. In order to overcome this anxiety, and to maintainproductivity and efficiency as required by mainstream discourse, self-control and self-discipline mechanisms that intensify the neoliberal subjectivity are put to work. As a result, related to destructive emotions such as de-territorialization, inadequacy and burnout, intense individualization and self-alienation emerge, to the point of us asking questions such as 'Did I want this?' or 'Is this my life?'. These emotions and the effects of neoliberal subjectivity are also the basis on which the dissemination, legitimization and internalization of flexible capitalist exploitation lies.

This study also highlights the key role that the viability of opting out of digital technologies and digital unavailability played during the pandemic. The more enforcement of the digital increases, the more digital disengagement becomes impossible. Understanding and explaining such a relationship may be a starting point for a discussion. However, the permanency of the damage apart from the pandemic, related to work–life imbalance and digital engagement should also be considered. The viability of opting out of digital technologies as a right, especially by creative workers, should be defended against the flexible capitalist exploitation disseminated in every aspect of our lives.

Notes

[1] According to the findings of research on Turkey's cultural and creative industries during the pandemic, the arts and culture sector was not adequately supported by the government; although partly supported by the local authorities, professional associations and unions during the pandemic, they were not sufficient to influence the policy makers to provide adequate support (Özarslan, 2021). The study also revealed that, in order to ensure the long-term sustainability of the industry, urgent public support, applicable policies and new business models are needed (Özarslan, 2021).

[2] An academic job may also be framed as cultural/creative work considering the short-term contracts, flexible and extra voluntary working hours, autonomous, individualistic and precarious work, affective labour, motivation of love, the requirement for self-discipline, self-control and thinking outside the box, creative problem-solving skills, and other reasons that I have tried elaborate in subsequent sections (see also Gill and Pratt, 2008; Vatansever and Gezici-Yalçın, 2015).

[3] By 'all in one' workers, I mean those who can do a little bit of every job when necessary, offer a comprehensive package with versatile competencies and skills, and are also open to flexibility in job descriptions when needed; in other words, flexibility in qualifications and work (Standing, 2015).

[4] During the pandemic, digital tools came up with new features for this digital engagement (for example, Slack added a daily reminder to post standup in channel).

[5] Approximately five months of the period I worked on social media coincided with the system of working from home during the pandemic. When I started to work as an assistant professor at a foundation university towards the end of 2020, we were expected to work in the office (except

for temporary periods when the state stipulated rotating work), even though the risks posed by the pandemic continued and the vaccine had not yet been found.

[6] In my social media job, I was also one of those who produce such motivational social media content.

[7] Another study on creative workers (Scolere et al, 2018) also outlined that the belief of a pay off – whether a new connection or collaboration – demands an investment in the form of time and energy, as part of the entrepreneurial self, framed within a larger economy that privileges certain positions.

[8] I think this depth comes from looking at my wounds and flaws by embracing them but as if they were not mine. Even though this journey filled with existential crises was a more tiring process than looking at, listening to and analysing the experiences of others, I tried to maintain self-reflexivity, integrity and transparency.

References

Ashton, D. (2011) 'Upgrading the self: technology and the self in the digital games perpetual innovation economy', *Convergence*, 17(3): 307–321.

Bagger, C. and Lomborg, S. (2021) 'Overcoming forced disconnection: disentangling the professional and the personal In pandemic times', in A. Chia, A. Jorge and T. Karppi (eds) *Reckoning with Social Media*, London: Rowman & Littlefield, pp 167–186.

Baudrillard, J. (2012) *Tüketim Toplumu [The Consumer Society]*, translated by H. Deliceçaylı and F. Keskin, Istanbul: Ayrıntı.

Bauman, Z. (2017) *Akışkan Modernite [Liquid Modernity]*, translated by S.O. Çavu, Istanbul: Can Yayınları.

Bourdieu, P. (2017) 'Prekaryalaşma Bugün Her Yerde' in *Contre-feux: Propos pour servir à la résistance contre l'invasion néo-libérale [Karşı Ateşler-I Neoliberal İstilaya Karşı Direnişe Hizmet Edecek Sözler]*, Işık Ergüden (ed), Istanbul: Sel Yayıncılık, p 113.

Denzin, N.K. (1997) *Interpretive Ethnography: Ethnographic Practices for the 21st Century*, Thousand Oaks, CA: Sage.

Denzin, N.K. (2003) *Performative Ethnography: Critical Pedagogy and the Politics of Culture*, Thousand Oaks, CA: Sage.

Duffy, E.D. (2016) 'The romance of work: gender and aspirational labour in the digital culture industries', *International Journal of Cultural Studies*, 19(4): 441–457.

Ellis, C. and Bochner, A.P. (2000) 'Autoethnography, personal narrative, and personal reflexivity', in N.K. Denzin and Y.S. Lincoln (eds) *The Sage Handbook of Qualitative Research*, Thousand Oaks, CA: Sage, pp 733–768.

Fuchs, C. (2015) *Dijital Emek ve Karl Marx [Digital Labor and Karl Marx]*, translated by T.E. Kalaycı and S. Oğuz, Ankara, Turkey: Notabene.

Gershon, I. (2011) 'Un-friend my heart: Facebook, promiscuity, and heartbreak in a neoliberal age', *Anthropological Quarterly*, 84(4): 865–894.

Gill, R. and Pratt, A. (2008) 'Precarity and cultural work in the social factory? Immaterial labor, precariousness and cultural work', *Theory, Culture and Society*, 25(7–8), 1–30.

Gorz, A. (1995) *İktisadi Aklın Eleştirisi [Critique of Economic Reason]*, translated by I. Ergüden, Istanbul: Ayrıntı.

Gregg, M. (2011) *Work's Intimacy*, Cambridge, UK: Polity Press.

Haiven, M. (2016) *Hayali Sermaye [Cultures of Financialization]*, translated by Y.E. Kara, Istanbul: Koç Üniversitesi Yayınları.

Hesmondhalgh, D. and Baker, S. (2011) *Creative Labour: Media Work in Three Cultural Industries*, London: Routledge.

Hesselberth, P. (2enaturalizinges on enaturalizing and the right to disconnect', *New Media & Society*, 20(5): 1994–2010.

Holman Jones, S. (2005) 'Autoethnography: making the personal political', in N.K. Denzin and Y.S. Lincoln (eds) *The Sage Handbook of Qualitative Research*, Thousand Oaks, CA: Sage, pp 763–792.

John, N. and Dvir-Gvirsman, S. (2015) '"I don't like you anymore': Facebook unfriending by Israelis during the Israel–Gaza conflict of 2014', *Journal of Communication*, 65(6): 953–974.

Kuntsman, A. (2012) 'Introduction', in A. Karatzogianni and A. Kuntsman (eds), *Digital Cultures and the Politics of Emotion*, London: Palgrave Macmillan, pp 1–20.

Kuntsman, A. and Miyake, E. (2019) 'The paradox and continuum of dienaturalizingement: enaturalizing digital sociality and technological connectivity', *Media Culture & Society*, 41(6): 1–13.

Lee, H.K. (2017) 'The political economy of 'creative industries', *Media, Culture & Society*, 39(7): 1–11.

Mills, C. W. (1951) *White Collar: The American Middle Classes*, New York: Oxford University Press.

Özarslan, Z. (2021). 'Yaratıcı ve Kültürel Endüstriler ve COVID-19 Pandemi Döneminde Türkiye'de Kültür ve Sanat Sektörlerinin Durumu', *Alternatif Politika*, 13(2): 371–408 [abstract in English].

Scolere, L., Pruchniewska, U. and Duffy, B.E. (2018) 'Constructing the platform-specific self-brand: the labor of social media promotion', *Social Media + Society*, 4(3): https://doi.org/10.1177/2056305118784768.

Sennett, R. (2013) *Karakter Aşınması [The Corrosion of Character]*, translated by B. Yıldırım, Istanbul: Ayrıntı.

Seymour, K., Skattebol, J. and Pook, B. (2020) 'Compounding education disengagement: COVID-19 lockdown, the digital divide and wrap-around services', *Journal of Children's Services* 15(4): 243–251.

Sezgin, S. (2020) *Dijital Oyun Ekosistemi: Yaratıcı Endüstri ve Emek*. Ankara, Turkey: Alternatif Bilişim.

Standing, G. (2015) *Prekarya [The Precariat]*, translated by E. Bulut, Istanbul: İletişim.

Timeturk (2020) 'Erdoğan'dan koronavirüs mesajı: 2. Dünya Savaşı'ndan sonraki en büyük kriz', *Timeturk* [online] 21 April. Available from: https://www.timeturk.com/erdogan-dan-koronavirus-mesaji-2-dunya-savasi-ndan-sonraki-en-buyuk-kriz/haber-1444746 [Accessed: 29 March 2023].

Vatansever, A. and Gezici Yalçın, M. (2015) *Ne Ders Olsa Veririz: Akademisyenin Vasıfsız İşçiye Dönüşümü*, Istanbul: İletişim.

Virilio, P. (1998) *Hız ve Politika [Speed and Politics]*, translated by M. Cansever, Istanbul: Metis.

THREE

'#RoeVsWadeOverturned: Any Idea How Fast Your #PeriodtrackingApp Can Lead To Jail?': Digital Disengagement and the Repeal of *Roe vs Wade*

Sam Martin

Introduction

Over the last decade, there has been a large increase in people's use of mobile health apps (mHealth) to track not only their general health, fitness levels and mental health (Ming et al, 2020), but also more specific medical conditions and biological indicators, such as chronic illnesses, obesity, and fertility and menstruation cycles (Byambasuren et al, 2018; Polzer et al, 2022). While the use of mHealth apps was indeed prevalent before the pandemic, during the pandemic itself, the increased reliance on technology for virtual communication, data collecting and data sharing during lockdowns, quarantines and beyond has meant that people have focused more on using technology and mHealth apps than ever before. In this chapter, 'pre-pandemic' refers to the 10-year period between 2010 and the start of the COVID-19 pandemic in February 2020, 'during the pandemic' refers to the period from February 2020 until the first COVID-19 vaccine rollouts in

December 2021 to March 2022, and 'post-pandemic' refers to the period between the time when the COVID-19 vaccine programme became more globally established and worldwide lockdowns decreased (from April 2022 to September 2022 and beyond). The chapter discusses the continuation of use of period apps during the pandemic (from February 2020 to March 2022), and the worry expressed by women regarding use of the apps that they had used before and during the pandemic during the window of time after *Roe vs Wade* was repealed in June 2022, with a continued discussion of app surveillance and a desire to digitally disengage from or circumvent this surveillance in the months going forward.

During the pandemic, one of the main focuses of mHealth apps was to develop an understanding of how to navigate the coronavirus during the pandemic and understand how it affected our bodies (Shah et al, 2021), not only in terms of initial infection and recovery (or prolonged symptoms/non-recovery in the context of long COVID), but also in terms of the effect of this new virus on everyday bodily functions, such sleep, mental health and mobility (Robertson et al, 2021). For women, there was a focus on the use of mHealth apps to understand how COVID-19, or indeed the COVID-19 vaccine(s), might affect their menstrual cycles in terms of length, heavier/lighter menstrual flow or intermenstrual spotting (Edelman et al, 2022), pregnancy or fertility levels (Human Health, 2021). Menstrual and fertility tracking apps (MFTAs) (Polzer et al, 2022) were already used quite heavily before the pandemic, with over 100 million people using them by 2020 (Gambier-Ross et al, 2018; Fowler et al, 2020). MFTAs also became one of the ways that women found to monitor their cycles without the need for direct consultation with a doctor at the height of the pandemic, when health services were stretched and access was restricted (Winter and Davidson, 2022). While the increase in reliance on these apps were seen by some as a welcome use of technology (Dodgson, 2020), others worried about the accuracy of the data shared by these apps in terms of helping women conceive or avoid unexpected pregnancies without the guidance of a doctor (Earle et al, 2020). A strong body of literature has also evolved in the fields of self-tracking and the quantified self, in which questions in relation to the tracking and surveillance of female bodies have started to emerge (Karlsson, 2021; Siapka and Biasin, 2021). The rise of MFTAs may also be situated in relation to ecofeminism and natural fertility education, in which discussions on reclaiming of bodily autonomy have given rise to a plurality of discourses about natural family planning and the contemporary digitalization of fertility awareness (Fotopoulou, 2016), which – within the context of the enforced digital surveillance of MFTAs – further complicates the implications of the aftermath of *Roe vs Wade* on female bodily autonomy.

This chapter aims to use the case of MFTAs as a starting point to explore broader issues surrounding the sharing, surveillance, and, in some cases, criminalization of data with or by third parties and authorities via these platforms. What impact will this have on the enactment of public health policies, and the need to adapt to the sometimes-negative effects of public health apps on the lives of individuals and business infrastructure going forwards? This chapter proactively engages with the nuances of digital opt-out and disengagement, and address the ongoing tensions that can arise between a more technologically informed public who are aware of their right to opt-out and digitally disengage, and the need for governments to protect the public and their right to self-track their health without harsh, infringing and punitive regulation.

An additional worry with regard to data shared by women within MFTAs both before and during the pandemic is the question of digital surveillance and surveillance capitalism. The detailed monitoring and datafication of women's bodies in relation to hormones, menstrual cycle and fertility has led to increased calls for better protection of women's rights and data, especially when viewed from the perspective and conditions of surveillance capitalism (Ford et al, 2021; Amelang, 2022).

Previous literature has also suggested that users actively engage in opting out of sharing their data within these apps, with prompts to digitally disengage with MFTAs that pose a threat to body autonomy and privacy (Karlsson, 2021). However, both before and during the pandemic, there have been an increasing number of reports about data about women's menstrual/fertility cycles being used to criminalize them, first between 2010 and 2019, to prosecute women for miscarriages (Das, 2022), and then during the pandemic, where, in one case, data from specific MFTAs such as the Flo app (Merken, 2021) were handed over to prosecuting authorities when the app company was subpoenaed by a state that was prosecuting a woman for unexplained miscarriages (Das, 2022; Elliott, 2022a). It is arguable that this use of women's bodily data to criminalize them uncovers a shift from surveillance capitalism into a form of digital authoritarianism, where personal data are used to criminalize people using a tool that has been advertised to help them manage their health. Alternatively, one may also argue that MFTAs (and perhaps even social media data) now co-exist with a dual form of surveillance capitalism and digital surveillance authoritarianism, especially as app companies can both profit from and criminalize women at the same time based on the datafication of their bodies. While this dual form of digital surveillance and criminalization of biological data is already in itself problematic, the most recent repeal of pro-abortion rights in the US in June 2022 (Garamvolgyi, 2022) has made the argument for scrutiny of how MFTAs use and share women's data with authorities and third parties even more important in terms of human and bodily rights, whereby various acts of digital disengagement may be needed to protect personal and health-related freedoms.

In June 2022, the US Supreme Court officially overturned the protection that the case *Roe vs Wade* had given women in terms of their right to have an abortion and the reproductive health rights in relation to this (Garamvolgyi, 2022). In the lead up to and immediately after this repeal, women in both the US and globally have deleted period-tracking apps from their smartphones, due to fears that the data collected by them may be used against them in any future criminal cases in states/countries where abortion has become illegal (Garamvolgyi, 2022; Poli, 2022). This form of mass digital disengagement enacted by tens of thousands of women after the repeal of *Roe vs Wade* shows just how embedded the datafication of bodies has become in our everyday lives, and how acutely aware women are becoming with regard to how vulnerable this has made them in relation to state surveillance capitalism. It is also suggested that this form of collective digital disengagement falls within the context of digital disengagement as a form of activism or protest (Min, 2019), rather than disillusionment or burnout towards technology use. Questions also arise regarding how various kinds of digitally tracked knowledge production are 'bodily anchored and situated' (Ford et al, 2021), and, more currently within the context of the overturning of *Roe vs Wade*, how this data, shared among many data providers, may in turn be shared with more conservative and punitive authorities to restrict women's bodily autonomy in relation to questions of abortion and reproductive rights.

This chapter uses social media discourse and sentiment analysis to understand how – in the immediate aftermath of the overturning of *Roe vs Wade* – women and their allies strove to effectively delete data, digitally disengage, algorithmically disrupt or opt-out of sharing their data via period-tracking apps. I also look at the hacks, instructions and hashtags used to share information and guide others. In the context of an increased reliance on mHealth apps during the pandemic, analysis of the discourse on social media shows that women actively want to continue to use these apps to help them manage their period and fertility cycles; however, there is a strong desire to disengage from the threat of being prosecuted should they need an abortion while doing so. Thus through social media analysis, this chapter looks to understand the tensions between digital opt-out and disengagement from the point of view of users aware of their right to opt-out and digitally disengage in a politically complex situation, and avoid the use of their health data being used to as a punitive tool.

Roe vs Wade and period-tracking: data privacy versus digital disengagement

The most interesting aspects of the social media response to the repeal of *Roe vs Wade* are how apps are used to spy on and limit women's bodily autonomy within the context of access to healthcare related to abortion or miscarriage, and the strategies of digital disengagement that can be used as resistance. I use text and discourse analysis to investigate the various levels of digital spying and sharing of women's data through MFTAs, specifically by analysing the social media discussion shared by users regarding their scrutiny of the terms and conditions of MFTAs and their methods of data collection. Various levels of digital disengagement are also explored through a discourse analysis of hacks, opt-outs and tips that users have shared on social media in terms of how to disengage via permanent deletion of apps and movement to analogue forms of tracking or movement away from harmful apps that sell data to authorities to those that are more trustworthy in terms of protection of data from anti-abortion lawmakers.

In order to capture this range of discussion, social media data were collected using the social media monitoring software Brandwatch™ (https://www.brandwatch.com) between 1 May and 7 August 2022. In the first stage, an initial horizon search of social media using search terms related to *Roe vs Wade* and tracking apps (Appendix 3.1) produced 7012 mentions; however, this search returned posts that were unconnected to *Roe vs Wade* in relation to period trackers, so a second Boolean search term was devised using an advanced Boolean search (Appendix 1), that incorporated all mentions of *Roe vs Wade*, discussion around period app privacy, deletion of apps to protect data, and mentions of specific apps that were trending. The second

Boolean search term returned a total of 3061 relevant mentions by 2286 unique authors, with a peaks of 130 mentions on 4 May, after leaked rumours on 2 May (Garamvolgyi, 2022) that US Supreme Court was going to repeal *Roe vs Wade*, and then a peak of 399 mentions between the day of repeal on 24 June and American Independence Day on 4 July 2022. Overall, most social media discussion occurred on Twitter (80 per cent), followed by Tumblr discussions (9 per cent), Reddit discussions (4 per cent), blog articles (4 per cent), forum discussions (2 per cent) and discussions in blog responses that mentioned period-trackers and *Roe vs Wade* (1 per cent).

Using a sentiment coding framework to look at posts specifically within the context of *Roe vs Wade* and period-tracking apps (Appendix 3.2), an analysis of all keywords shared in the dataset found that, while the majority of information shared was news articles, with no human comment (74 per cent), any mentions with human comment (such as quotes or retweets) were overall negative (19 per cent), with just 7 per cent of mentions being positive. The majority of topics of concern were linked to issues of privacy in relation to period-tracking apps, fears and worry about users being 'outed' if tech companies were subpoenaed for information by anti-abortion law makers, worry about levels of tracking, hashtag campaigns for abortion rights and women's rights, as well as discussions of the politics behind the repeal of *Roe vs Wade*, discussion about deleting apps, and finally discussion about what specifically might be done with personal and private data. Most emojis shared were linked to scepticism, anger and crying, while more positive emojis shared were the dove emoji (for peace) and the love smiley face emoji (in response to efforts by some tech companies to safeguard data). Examples of positive terms and hashtags that were shared were those in relation to pro-choice in support of women's rights to have an abortion (#prochoice #mybodymychoice), those promoting resistance to companies sharing data (#resistance), and discussion with regard to voting for the Democratic party in the November 2022 elections in the hope of changing the decision (#votethemout #votebluein22 #bluevoices), while other hashtags pushed explicitly for digital disengagement altogether, in terms of using paper instead of technology to track periods (#analogistheanswer).

The question of how safe period-tracking apps are for women in anti-abortion states to use became more prominent in the weeks and months after the repeal of *Roe vs Wade*, whereby, under more increased scrutiny, those apps that had pledged to protect users' data saw a surge of new users, while those who were more evasive about how data was shared with third parties or protected overall suffered considerable decreases in the number of users willing to share their data (Poli, 2022). By mid-July, popular news outlets also became engaged with the discussion of digitally disengaging from the use of period-tracking apps, and evasion of surveillance, publishing articles such as those listing the most popular period-tracking apps ranked by data privacy (Poli, 2022). In a similar way to academic papers that discussed the use of digital opt-out toolkits to understand how apps share data, and enact agency by opting out of using those that did not offer good protection (Kuntsman et al, 2019), in some media articles, journalists and users also shared similar toolkits with similar levels of technical detail, with the aim of quickly educating women as to what types of data can be misappropriated by apps (location, name, period cycles), the degree with which apps share information with third parties, and the degree to which the app allows a user to delete their personal data. Each app was then given a data privacy score with regard to how safe it is for women to use within an anti-abortion environment.

Demographics: digital disengagement versus algorithmic activism

In terms of demographics, there was a fairly even spread of both women and men discussing *Roe vs Wade* in the context of period trackers, with 58 per cent identifying as women versus 42 per cent identifying as men. While these data only focused on those that indicated their gender in their social media bios (35 per cent of the whole dataset), this shows an interesting balance in terms of gender discussion. When looking specifically for gendered hashtags such as #mybodymychoice (used by 40 per cent of women) versus #herbodyherchoice (used by 80 per cent of men) within the dataset, it was found that, specifically in posts with the hashtag #herbodyherchoice, men were engaged in discussions of algorithmic disengagement, by encouraging women to delete period-tracking apps. However, alongside these levels of digital disengagement, were calls for algorithmic activism (Min, 2019) that discussed disruption of the data within apps that were deemed untrustworthy. Algorithmic activism in this sense refers to activists who rally other users in an attempt to challenge and disrupt the algorithms by submission of fake or misleading data to social media platforms or apps so that the goal of that platform or app (for example, commodifying or sharing data with third parties) is thwarted (Min, 2019). Indeed, within the 2022 dataset analysed, it was found that men and women were also encouraging each other to engage in specific forms of algorithmic activism, such as by sharing posts encouraging men to inundate apps with false data that were actively selling women's data. An example of this was a post calling all men to register fake accounts with fake data, impersonating women with fake personas, so that anti-abortion law makers who obtained data from these apps would be led to believe false data.

In terms of trending topics between 1 May and 7 August, at the start of May, when rumours that the Supreme Court might repeal *Roe vs Wade* were leaked, there were more general discussions of the use of period-tracking apps to address notions of stigma around periods, abortion and reproductive rights in general, and opposition to the overturning of *Roe vs Wade*. However, these trending topics soon faded during June and

August 2022, after *Roe vs Wade* was overturned, and the more general topics was replaced with more focused discussion with regard to apps selling personal data linked to periods and birth control, as well as what law enforcement might do with the data, and methods such as end-to-end encryption that might be a way to evade data collection.

Geographical analysis of this discussion also found that topics connected to *Roe vs Wade* and digital disengagement were in fact discussed worldwide, despite this being an American law, affecting women's reproductive rights within US states. What is interesting here is that not only were citizens of countries with similar Western cultures such as the UK, Canada, Australia, Germany, the Netherlands and France engaged in discussions of *Roe vs Wade* and period trackers and issues of privacy, but that citizens of non-Western countries such as India and Malaysia were also worried about how governments who look to the US as an example may also change their stance towards women's reproductive rights and their access and utilization of private data. A deeper analysis of how the UK and other countries in Europe discussed this topic also found many references to the European General Data Protection Regulation) (Zaeem and Barber, 2018), and whether this might offer protection to women who were using American apps but perceived that their data privacy was protected under European rules. In contrast, American women discussed whether their data could be better protected by using European apps versus American apps whose developers might be easily subpoenaed to give up fertility or abortion-related data.

Overturning digital disengagement: claims of transparency versus digital activism

Some of the more popular period-tracking apps, such as Stardust realized that they needed to protect their market share by making it more apparent how much they protected women's data privacy, and took to social media to inform users of this, for example by using TikTok to share an announcement reassuring users about an end-to-end encryption update for all data shared on its app, stating that, because data were encrypted, if subpoenaed by a court, Stardust could not be compelled to hand over users' personal data (Stardust, 2022). The Stardust TikTok announcement was called out in the comments on their post and by other TikTok users who further investigated Stardust's terms of services documentation on their main website, and shared that information in comments under the post. Using the text analytics software, InfraNodus Paranyushkin (2019) performed a discourse analysis of the 5000 or so comments under the video, and a sentiment analysis around this discussion, and showed 86 per cent negative sentiment versus 14 per cent positive sentiment in terms of whether the Stardust TikTok post was legitimate in claiming that its encryption was secure versus claims by other users that this was false information, and that users should be cautious about using the app in a state that bans abortion. Discourse analysis also showed that the main topics of discussion in the comments on this video were in relation to deletion of apps that had already been downloaded, how to make sure that personal information was not shared with the government, specific mentions of deleting rival apps that had also been rumoured to sell women's data, whether apps could be believed when they stated they would not sell data, and encouraging other users to read terms of services related to period-tracking apps rather than going on the face value of shared videos. These topics of concern reveal a considerable amount of scepticism regarding the supposedly robust protection of privacy by some period-tracking apps, as well as some localized knowledge shared between users as to how to circumvent, avoid or evade personal data being sold onto third parties.

Another article (Cole, 2022) that undertook an investigation into the data privacy and security of users' data on period-tracking apps found that, despite their TikTok announcement, Stardust had forgotten to be clear about a loophole in their privacy policy, which actually stated:

> We may disclose your anonymized, encrypted information to third parties in order to protect the legal rights, safety, and security of the Company and the users of our Services; enforce our Terms of Service; prevent fraud; and comply with or respond to law enforcement or a legal process or a request for cooperation by a government or other entity, whether or not legally required.

Because of this discrepancy in telling users their data was safe, while also signalling to authorities that they would hand over users' data if asked, *Wired* (Elliott, 2022b) gave Stardust a lower safety rating. Following continued requests for comment, Stardust later updated its policy, to say that it would 'comply with or respond to law enforcement or a legal process or a request for cooperation by a government or other entity, when legally required. Any Health Data that the Company is legally required to share cannot be linked to [users] and will remain anonymous'.

It emphasized that its use of an 'encrypted wall' between users' personally identifiable information (email/phone/apple ID and so on) and the data that the users add to the Stardust app meant that, even if compelled to hand over data to authorities, the company actually would not have anything revealing to give them. However, some users remain sceptical about this, arguing that because of the selling of app data to not immediately declared/hidden parties, such as the likes of Facebook and Google/Alphabet by Stardust, which has already been prosecuted for doing so (Cole, 2022), the only really safe way around this is to be able to use the app anonymously, without an account, to avoid collection of any identifiable data.

The fact that such data can be subpoenaed in the wake of the repeal of *Roe vs Wade* is just a slightly newer twist on a practice that has been increasing in recent years, whereby in both the US and the UK, women coming in for medical treatment are being arrested for inducing an abortion. In the US, a recent case is that of Latice Fisher, a woman from Mississippi, who arrived at a hospital suffering from a miscarriage: 'she was later indicted on a charge of second-degree murder. Prosecutors used her search history and her online purchase of the abortifacient drug misoprostol as evidence to allege that she had murdered her foetus' (Elliott, 2022a).

This story is just the tip of the iceberg, however; with the National Advocates of Pregnant Women, a pro-choice advocacy group in the US, having recorded 1600 similar cases between 1973 and 2000, with approximately 1200 occurring in the last 15 years alone. In some of these cases, women were arrested for things such as 'falling down, or giving birth at home, [while] the vast majority involved drugs, and women of colour were overrepresented' (Levinson-King, 2021). There have been a total of 67 cases in the last 10 years in England and Wales, where, despite the 1967 Abortion Act, which legalized abortion as a healthcare right, a draconian 1861 Offences Against the Person Act, which deems it 'unlawful to procure a miscarriage using "poison", "an instrument" or "other means whatsoever"', has been used to prosecute vulnerable women who need help and support instead of the trauma of criminalization and jail time. One recent example is a 2021 case, in which 'a 15-year-old girl who had an unexplained early stillbirth was subjected to a year-long criminal investigation that saw her text messages and search history examined. Police dropped the case after a coroner concluded the pregnancy ended because of natural causes' (Das, 2022).

It is arguable that the app-ization and datafication of health adds another, newer twist in an already existing and old form of systematic reproductive violence.

Conclusion

It is arguable that both before and during the pandemic, the increased reliance on the use of mHealth apps (Winter and Davidson, 2022), the gradual naturalization of the habit of increased reliance on data to help us manage our lives/health data, has created a dangerous precedent, whereby users have little knowledge of how and where their data are shared, and even less control over how these data may be used to criminalize them. This mirrors other practices of rapid reliance on digital monitoring of health data during the COVID-19 pandemic, in the form of contact-tracing apps and COVID vaccine passports, whereby this gradual but heavy reliance on digital mapping of bodies across local and national borders has also led to questions of digital health surveillance and the rights of people to digitally disengage from tracking their bodies by governments and states (Cofone, 2020). In these cases, forms of digital disengagement were enacted by deletion of such apps, or in some cases the refusal to travel until vaccine restrictions and the requirement for contact tracing were relaxed (Gstrein et al, 2021). Within the context of period-tracking apps, the stark reality of the way that this can escalate into authoritarian surveillance in the context of pregnant bodies in the wake of the repeal of *Roe vs Wade*, makes the need for more regulation of the misuse of personal data by the state even more urgent. Indeed, the mass harvesting of data linked to people's bodies and their movements in relation to this may arguably fuel development in after-the-fact datafication, whereby digital health data has the potential to be mis-used for other purposes not consented to by the original app user (Kuntsman et al, 2019). Such after-the-fact datafication in terms of the selling of health data to private companies for profit or non-consenting research continues to be a contention, with calls for stronger regulation of these practices continuing today (Stenström, 2023).

In the interim, as COVID-19 is far from over, and future pandemics are becoming more probable (Shadbolt et al, 2022), and indeed everyday life is more heavily impacted by the use of mHealth apps, how may current forms of digital disengagement help safeguard user autonomy? Overall, the study described in this chapter has found that, in times of upheaval regarding various aspects of human rights, most specifically women's rights in the context of the repeal of *Roe vs Wade*, the concept of digital disengagement can be utilized as a form of digital resistance, protest and protection in a fight for stronger rights to privacy and bodily autonomy, especially in terms of the datafication of personal and body-centred data that have been turned from a quantified self-help tool to a means to punish women seeking healthcare. In this case, however, digital disengagement has been played out to varying degrees, from absolute deletion of all period-tracking apps, deletion of one app in favour of apps that gather and utilize data in less harmful ways and protect data privacy, to use of social media to enact a form of digitally engaged algorithmic activism by flooding harmful apps with misleading data.

What is interesting in the context of those deleting harmful apps in favour of less harmful ones is the decision of users not to completely disengage from using all apps, but to gain agency in the way that certain apps are used. This in turn has been helped by liberal media and tech agencies, who – perhaps in the context of a mixture of understanding the need of women to use period apps, as well as the need to continue capitalist surveillance in a less punitive context – have also worked to create toolkits to help women understand more about how their data are shared, and what degree of 'safety' each period-tracking app does or does not afford them. However, like contact-tracing apps, period-tracking apps are not the only ways that technology can be used to connect women or their allies to abortion, this can also be done via games, adverts and other technologies that track and share activity linked to apps installed on smartphones with third parties. It is recommended that any app that collects sensitive information about a person's health/body should be given additional scrutiny, and that lessons learned from concern about issues regarding tracking of bodily data in the aftermath of the repeal of *Roe vs Wade* be applied to all apps. This, together with the need to look at what apps do in terms of their privacy policies and terms of service, brings us back to the need to push for additional and cleaner measures to allow users to more effectively opt-out of their personal data being tracked and shared. While deletion of problematic apps may be the start, being able to effectively, and retroactively, delete all previously collected data is also of key importance, especially when human rights and bodily autonomy are at stake (Kuntsman et al, 2019). It may be that, in some cases, where punitive and legal repercussions put personal health data at risk, perhaps the analogue method using paper calendars or diaries is the safer way to completely digitally disengage.

Digital disengagement as a form of resistance may be one of the stronger tools to use to gain back control over how one's digital health data is used, especially as, without constant monitoring of changes to terms and conditions updated by app companies in relation to how personal rights are protected, it can be difficult for a general user to stay on top how their data is being (mis)managed. On the other hand, where one cannot fully disengage, a balance of digital resistance and partial disengagement may be the route that users opt to take going forwards. This may take the form of mixing manual tracking via paper planners/journals with basic calendar apps that holding nothing more than coloured data for the topic being tracked (perhaps red/pink for menstruation days), or for mHealth apps, using free versions of apps without actually creating accounts that link to the user via email and other personal data. Indeed, post-pandemic, it will be interesting to see what forms these enactments of digital resistance to apps used during the pandemic evolve into, as the increased use of digital surveillance necessitates a balance between users taking control of their data versus the companies/authorities that may make harmful use of them.

Appendix 3.1: Boolean searches

Preliminary search: 7012 mentions

(periodtracker OR DeleteYourPeriodTrackingDevice OR deleteapp OR "delete app" OR "delete the app" OR deleteflo OR "Spot On app" OR SpotOnapp OR stardustapp OR "stardust app" OR deletestardust OR "delete stardust" OR clueapp OR deleteclue OR "delete clue" OR deleteclueapp) AND ("Roe vs wade" OR abortionban OR roevswade OR roevwade OR prochoice OR ProLife OR Resistance OR SupremeCourt OR AbortionRightsAreHumanRights OR womenrights OR menstruation OR GDPR OR abortionban OR abortionishealthcare OR waronwomen OR democracynotautocracy OR abortionrights)

Final search: 3,061 mentions (1 May to 7 August 2022, excluding news)

(periodtracker OR DeleteYourPeriodTrackingDevice OR deleteapp OR "delete app" OR "delete the app" OR deleteflo OR "Spot On" OR SpotOn OR stardust OR deletestardust OR "delete stardust") AND ("Roe vs wade" OR abortionban OR roevswade OR roevwade OR prochoice OR ProLife OR SupremeCourt OR AbortionRightsAreHumanRights OR womenrights OR menstruation OR GDPR OR abortionban OR abortionishealthcare)

Appendix 3.2: Sentiment analysis framework

Positive (P)

• Post communicates positive views towards status quo regarding protection of data privacy in period–tracking apps.

- Post encourages uptake in use of standard period-tracking apps.
- Post tackles negative comments regarding status quo in terms of data sharing and the use of period-tracking apps.

Negative (N)

- Post communicates negative/sceptical views towards status quo regarding protection of data privacy in period-tracking apps.
- Post discourages the use of standard period-tracking apps.
- Post encourages others to delete period-tracking apps to protect privacy.
- Post encourages others to disrupt the quality and authenticity of data shared to apps that do not advocate adequate protection of personal data.

Neutral (P)

- Post shares articles with neither positive nor negative comments towards the status quo regarding protection of data privacy in period-tracking apps.
- Post shares articles with neither positive nor negative comments towards uptake in use of standard period-tracking apps.
- Post shares articles with neither positive or negative comments regarding status quo in terms of data sharing and the use of period-tracking apps.

References

Amelang, K. (2022) '(Not) safe to use: insecurities in everyday data practices with period-tracking apps' in A. Hepp, J. Jarke and L. Kramp (eds) *New Perspectives in Critical Data Studies*, Basingstoke, UK: Palgrave Macmillan, pp 297–321.

Byambasuren, O., Sanders, S., Beller, E. and Glasziou, P. (2018) 'Prescribable mHealth apps identified from an overview of systematic reviews', *NPJ Digital Medicine* 1(1): 1–12.

Cofone, I. (2020) 'Immunity passports and contact tracing surveillance', *Stanford Technology Law Review*, 24: 176.

Cole, S. (2022) 'The #1 period tracker on the app store will hand over data without a warrant', *Vice* [online] 27 June. Available from: https://www.vice.com/en/article/y3pgvg/the-1-period-tracker-on-the-app-store-will-hand-over-data-without-a-warrant [Accessed 1 August 2022].

Das, S. (2022) 'Women accused of illegal abortions in England and Wales after miscarriages and stillbirths', *The Guardian* [online] 2nd July. Available from: https://www.theguardian.com/world/2022/jul/02/women-accused-of-abortions-in-england-and-wales-after-miscarriages-and-stillbirths [Accessed 30 August 2022].

Dodgson, L. (2020) 'The entrepreneur who coined the term "FemTech" founded a period tracking app that's helping women understand and accept their bodies', *Insider* [online] 5 June. Available from: https://www.insider.com/founder-of-clue-ida-tin-coined-the-term-femtech-2020-6 [Accessed 1 August 2022].

Earle, S., Marston, H.R., Hadley, R. and Banks, D. (2021) 'Use of menstruation and fertility app trackers: a scoping review of the evidence', *BMJ Sexual & Reproductive Health*, 47: 90–101.

Edelman, A., Boniface, E.R., Male, V., Cameron, S.T., Benhar, E. and Han, L., et al (2022) 'Association between menstrual cycle length and covid-19 vaccination: global, retrospective cohort study of prospectively collected data', *BMJ Medicine*, 1(1), doi: 10.1136/bmjmed-2022-000297.

Elliott, V. (2022a) 'The fall of 'Roe' would put big tech in a bind', *Wired* [online] 6 May. Available from: https://www.wired.com/story/big-tech-roe-abortion/ [Accessed 30 May 2022].

Elliott, V. (2022b) 'Fertility and period apps can be weaponized in a post-Roe world', *Wired* [online] 7 June. Available from: https://www.wired.com/story/fertility-data-weaponized/ [Accessed 20 June 2022].

Ford, A., de Togni, G. and Miller, L. (2021) 'Hormonal health: period tracking apps, wellness, and self-management in the era of surveillance capitalism', *Engaging Science, Technology, and Society*, 7(1): 48–66.

Fotopoulou, A. (2016) 'From egg donation to fertility apps: feminist knowledge production and reproductive rights', in *Feminist Activism and Digital Networks*, London: Palgrave Macmillan, pp 91–121.

Fowler, L.R., Gillard, C. and Morain, S.R. (2020) 'Readability and accessibility of terms of service and privacy policies for menstruation-tracking smartphone applications', *Health Promotion Practice*, 21(5): 679–683.

Gambier-Ross, K., McLernon, D.J. and Morgan, H.M. (2018) 'A mixed methods exploratory study of women's relationships with and uses of fertility tracking apps', *Digital Health*, 4: 2055207618785077.

Garamvolgyi, F. (2022) 'Why US women are deleting their period tracking apps', *The Guardian* [online] 28 June. Available from: https://www.theguardian.com/world/2022/jun/28/why-us-woman-are-deleting-their-period-tracking-apps [Accessed 1 August 2022].

Gstrein, O.J., Kochenov, D. and Zwitter, A. (2021) 'A terrible great idea? COVID-19 'vaccination passports' in the spotlight'. Available from: https://www.compas.ox.ac.uk/2021/a-terrible-great-idea-covid-19-vaccination-passports-in-the-spotlight/ [Accessed 4 April 2023].

Human Health (2021) 'Why the covid-19 vaccines are an important women's issue', Available from: https://ignifi.co.uk/covid-19-vaccines-and-women/ [Accessed 1 June 2022].

Karlsson, A. (2021) 'Whose bodies? Approaching the quantified menstruating body through a feminist ethnography', in B. Ajana, J. Braga and B. Guidi (eds) *The Quantification of Bodies in Health: Multidisciplinary Perspectives*, Bingley, UK: Emerald Publishing Limited, pp 119–134.

Kuntsman, A., Miyake, E. and Martin, S. (2019) 'Re-thinking digital health: data, appisation and the (im)possibility of 'opting out'', *Digital Health*, 5: 2055207619880671.

Levinson-King, R. (2021) 'US women are being jailed for having miscarriages', *BBC News* [online] 12 November. Available from: https://www.bbc.co.uk/news/world-us-canada-59214544 [Accessed 31 August 2022].

Merken, S. (2021) 'Fertility app maker Flo Health faces consolidated privacy lawsuit', *Reuters* [online] 3 September. Available from: https://www.reuters.com/legal/litigation/fertility-app-maker-flo-health-faces-consolidated-privacy-lawsuit-2021-09-03/ [Accessed 1 Ausgust 2022].

Min, S.J. (2019) 'From algorithmic disengagement to algorithmic activism: charting social media users' responses to news filtering algorithms', *Telematics and Informatics*, 43: 101251.

Ming, L.C., Untong, N., Aliudin, N.A., Osili, N., Kifli, N. and Tan, C.S., et al (2020) 'Mobile health apps on COVID-19 launched in the early days of the pandemic: content analysis and review', *JMIR mHealth and uHealth*, 8(9): e19796.

Paranyushkin, D. (ed) (2019) 'InfraNodus: generating insight using text network analysis', in *The World Wide Web Conference*, Paris: Nodus Labs, pp 3584–3589.

Poli, K. (2022) 'The most popular period-tracking apps, ranked by data privacy', *Wired* [online] 28 July. Available from: https://www.wired.com/story/period-tracking-apps-flo-clue-stardust-ranked-data-privacy/ [Accessed 20 July 2022].

Polzer, J., Sui, A., Ge, K. and Cayen, L. (2022) 'Empowerment through participatory surveillance? Menstrual and fertility self-tracking apps as postfeminist biopedagogies', in J. Fellows, L. Smith (eds) *Gender, Sex, and Tech!: An Intersectional Feminist Guide*, Toronto: Canadian Scholars, pp 163–165.

Robertson, M., Duffy, F., Newman, E., Bravo, C.P., Ates, H.H. and Sharpe, H. (2021) 'Exploring changes in body image, eating and exercise during the COVID-19 lockdown: a UK survey', *Appetite*, 159: 105062.

Shadbolt, N., Brett, A., Chen, M., Marion, G., McKendrick, I.J. and Panovska-Griffiths, J., et al (2022) 'The challenges of data in future pandemics', *Epidemics*, 40: 100612.

Shah, M.D., Sumeh, A.S., Sheraz, M., Kavitha, M.S., Maran, B.A.V. and Rodrigues, K.F. (2021) 'A mini-review on the impact of COVID 19 on vital organs', *Biomedicine & Pharmacotherapy*, 143: 112158.

Siapka, A. and Biasin, E. (2021) 'Bleeding data: the case of fertility and menstruation tracking apps', *Internet Policy Review*, 10(4): 1–34.

Stardust (2022) 'Stardust Period Tracker', *TikTok* [online] 26 June 2022, Available from: https://www.tiktok.com/@stardust.app/video/7113638686075571502 [Accessed 1 August 2022].

Winter, J.S. and Davidson, E. (2022) 'Harmonizing regulatory regimes for the governance of patient-generated health data', *Telecommunications Policy*, 46(5): 102285.

Zaeem, R.N. and Barber, K.S. (2020) 'The effect of the GDPR on privacy policies: Recent progress and future promise', *ACM Transactions on Management Information Systems (TMIS)*, 12(1): 1–20.

FOUR

#SnailMailRevolution: The Networked Aesthetics of Pandemic Letter-Writing Campaigns

Chelsea Butkowski

Introduction

Amid the social distancing, doom-scrolling and mounting Zoom fatigue that accompanied the COVID-19 pandemic in the US, people reached for pens, paper and postage stamps. Despite their long legacy, handwritten letters are often discussed as a novelty when social connection seems just a text or video call away. Nonetheless, the pandemic brought on sympathy card shortages and think pieces on the benefits of sending letters for a pandemic-addled, computer-saturated sociality. Penpal programmes blossomed, and, according to a survey by the US Postal Service, people reported sending more personal mail than they had before (US Postal Service, 2021). Even as the pandemic unevenly altered online social connection for many (Nguyen et al, 2020), the proliferation of handwritten letters presented an alternative to digital modes of social interaction and a supposed escape from the connective commodification engendered in platform economies (van Dijck, 2013). After all, media scholars have long argued that analogue epistolary technologies invite intimacy and social presence because of their textualized and material qualities, not in spite of them (Milne, 2010).

Beyond the cultural salience of letters, however, the popularity of hashtags about letter writing also grew on the same social media platforms that handwritten letters circumvent. Letter-writing aesthetics trended on Instagram long before COVID-19, but, within the altered climates of pandemic digitalities, their growth takes on new meaning. People shared images of letters they received and others they planned to send using hashtags such as #SnailMailRevolution and #SendLove. More telling, however, are the dozens of popular letter-writing campaigns that sprung up to combat pandemic social isolation and engage political causes, encouraging participants to share their missives on Instagram. Campaign communications and news articles discuss letter writing as a means to disengage from the digital and reconnect in novel yet nostalgic ways, which, according to journalist Tanya Basu (2020) 'makes snail mail possibly more powerful than email or text'. *The New York Times* reported that letter writing is 'getting people through this time' (Danovich, 2020), while the founder of a viral pandemic penpal project explained, 'There is something about [letter writing] being disconnected from this immediacy of online that feels enchanting to people' (Cooper, 2021).

Kuntsman and Miyake (2019: p 902) describe digital disengagement, including acts of disconnection, detox, refusal and withdrawal from digital systems, as 'a continuum of practices, motivations, and effects' rather than a bifurcated distinction between engaged and disengaged. Pandemic letter writing and sharing probe the boundaries of this continuum, inspiring questions about how layered technological apparatuses and interactional techniques come together to facilitate letter writing as a performance of digital disengagement within pandemic-era social frameworks. The letters and accompanying Instagram posts suggest a networked aesthetics that engages online and offline audiences through nostalgic modes of seemingly disengaged social connection. However, when people choose to put disengaged practices on digital display, they also echo persistent social inequities inherent in online visibility and pandemic experiences, particularly as those with the time and resources to write letters selectively opt-in to document their own disengaged activities.

To examine the engaged disengagement of Instagrammable letter writing as a response to the COVID-19 pandemic, I conducted a textual analysis of three US-based pandemic letter-writing campaigns and their manifestations on Instagram. These were #LettersAgainstIsolation, an initiative to write anonymous letters to senior citizens who had been forced into isolation by pandemic social-distancing measures; #TheBigSend, a citizen-driven campaign to send voter encouragement letters in advance of the 2020 US election; and #Penpalooza, an international penpal programme that started amid lasting lockdowns. Drawing together hash-tagged Instagram posts, I set out to investigate how people visualize digitally disengaged pandemic communication practices through digital technologies. Ultimately, I argue that, while digital disengagement is typically characterized as a subtractive phenomenon, pandemic letter writing demonstrates the additive potential of performing disengaged practices online. However, these potentials do not exempt Instagrammable letter-writing campaigns from the well-established socio-political pitfalls of non-participation (Kaun and Treré, 2020).

Displaying digital disengagement

Digital disengagement connotes various situated forms of breakage or interruption in usage of digital technologies, whether imposed or intentional and temporary or permanent (Kuntsman and Miyake, 2019). While the COVID-19 pandemic required some people to become more digitally engaged than ever before, it also contributed to pre-existing conditions of compulsory and repressive disengagement caused by unequal accessibility and restrictive governance on a global scale (Kaun and Treré, 2020; Treré, 2021). Nevertheless, much of the existing scholarship on digital disengagement pertains to individuals and groups who actively choose not to use the digital technologies and platforms available to them (Jorge, 2019; Kuntsman and Miyake, 2019). Despite initial scholarly framings of self-chosen disengagement as a withdrawal from sociality or political participation, researchers have more recently begun to explore the political potentials of silence, invisibility and rejection enabled through active digital disengagement and interference efforts (Kaun and Treré, 2020). Just as disengagement disrupts usage of digital technologies, it can also disrupt existing, unfair governance systems as a strategy for political resistance or active non-participation through activities such as hacktivism or boycotting (Casemajor et al, 2015; Natale and Treré, 2020).

However, not all acts of disengagement are politically potent or even politically motivated. The choice to disengage, disconnect or detox from digital media may also be associated with socially desirable and potentially idealized qualities at the individual level (Kuntsman and Miyake, 2019; Kaun and Treré, 2020). In turn, displaying disengagement may also become a social performance or a gambit for personal control (Portwood-Stacer, 2012). In a study on Instagram and digital disengagement, Jorge (2019) found that disconnection from social media is culturally positioned as a means of tapping into aesthetics of authenticity and 'good taste'. Natale and Treré (2020: p 628) add that 'non-digital and offline media experiences are portrayed as somehow more genuine, purer, less toxic, and intrusive forms of engagement and sociality' than digital experiences. These characterizations highlight some of the challenges associated with disengagement initiatives, positioning them as processes of neoliberal social deliverance and identity construction.

The pandemic resulted in hundreds of thousands of deaths in the US amid heightened political, economic and racial tensions, all of which have been magnified, in part, through online modes of interaction. In the face of this tragedy, letter writing presents a nostalgic alternative to over-connection through digital modes of social interaction. Niemeyer and Keightley (2020: p 1641) describe nostalgia as 'a specific affective modality of engaging with the past', producing feelings through recollections of foregone personal memories and aesthetics. While nostalgia can unite groups and the public through this shared effect, nostalgic media technologies can also serve as a means for coping with changes to personal media ecologies, including the shifts in everyday modes, practices or frequencies of communication, such as those wrought by COVID-19 safety measures (Menke, 2017). Letter writing also does not require complete disconnection from digital forms of communication. Instead, it is a practice of digitally disengaged communication that can replace, augment or supplement facets of dominant digital modes. Within the situated platform vernacular and aesthetics of Instagram (Gibbs et al, 2015; Leaver et al, 2020), pandemic letters represent a choice to communicate offline, even if the results are ultimately displayed online. As Yuan (2021) argues, the relationship between so-called 'fast' digital media and 'slow' analogue media is not binary but instead fluid, socially constructed and ongoing as new relationships between media modalities continue to emerge. When viewing digital disengagement as a continuum (Kuntsman and Miyake, 2019), Instagrammable letter writing has the potential to be both engaged and disengaged.

The politics of pandemic letter writing in networked cultures

Whereas letter writing was among the primary modes of mediated personal communication prior to inventions such as the telephone and the internet, handwritten letters have become more of a novelty than a necessity in digitally networked cultures. This is not to say that mail is a rarity, as personal correspondence is just one letter genre among many, including transactional mail and form solicitations (Barton and Hall, 2000). Nevertheless, first-class, person-to-person correspondence mail has declined globally over the past two decades, largely as the result of e-substitution by digital technologies (US Postal Service, 2018). In the US, pre-existing stresses on mail systems were compounded when the pandemic hit, as businesses cut costs by reducing mailings and online shopping skyrocketed. Lasting financial losses and new leadership placed the US Postal Service in jeopardy, even with an approaching presidential election dependent on mail-in voting. Calls to 'save the USPS' accompanied journalists proclaiming that 'even sending mail that's not explicitly political has become an inherently political act' (Basu, 2020).

The diverging technological affordances of material and digital modes of correspondence structure their potentials for pandemic-era communication. Technological affordances describe how the features of a given technology shape the available or perceivable behaviours and social actions of its users (Bucher and Helmond, 2017). While digital technologies offer rapid and ephemeral exchanges mediated through sometimes depersonalizing digital interfaces, handwritten letters present a more gradual and tangible form of communication personally created by the sender. Letter writing has long been associated with intimacy, presence and authenticity; connotations that have deepened in the digital era (Milne, 2010). I position the choice to correspond via snail mail instead of digital means as a partial, if marginal, displacement of digital sociality. To better understand this choice and its digital representations on Instagram, I examine three US-based pandemic letter-writing campaigns to answer the following research question: How do campaign organizers and participants represent, aestheticize and even idealize disengaged practices of pandemic letter writing on Instagram?

Letters Against Isolation, The Big Send and Penpalooza are large-scale initiatives that originated and organized in online spaces, encouraging participants to write and mail letters during the COVID-19 pandemic. Letters Against Isolation sends handwritten letters to senior citizens amid pandemic social-distancing measures.[1] Started by teenage sisters to stay in touch with their isolated grandparents during early lockdowns in the US, the ongoing campaign has expanded to send hundreds of thousands of letters to assisted living facilities and senior care homes internationally. The Big Send was a get-out-the-vote letter-writing campaign for the 2020 US presidential election organized by the non-profit social movement organization Vote Forward.[2] Centring on the date when volunteers were encouraged to mail their letters, 17 October 2020, Vote Forward reported sending over 17 million letters to historically under-represented eligible voters in 21 states, resulting in a 0.8 percentage point increase in voter turnout. Penpalooza is a penpal exchange programme that originated from New Yorker journalist Rachel Syme's social media network in the summer of 2020.[3] The campaign operates through Elfster.com, a website that is designed for organizing 'Secret Santa' gift exchanges. Penpalooza has grown to connect over 10,000 penpals in 75 countries.

Research design

To study the networked aesthetics of these pandemic letter-writing campaigns on Instagram, I used MAXQDA to organize data and iteratively develop codes to perform a textual analysis of Instagram posts based on the hashtag associated with each campaign. I selected these three letter-writing campaigns by balancing their similarities, such as widespread popularity and digital components, with their divergent goals, designs and target audiences. For example, campaign hashtags varied in popularity and context during data collection in September 2021, and my sampling techniques varied accordingly. All posts were sampled manually through a process of screen-shotting photos and side matter such as caption text and comments, all of which were subject to analysis. Although the campaign hashtags varied in popularity (#LettersAgainstIsolation, 241 posts overall; #TheBigSend, 3123 posts on 17 October 2020;

#Penpalooza, 1186 posts overall), I narrowed each corpus through a systematic sampling process that weights them evenly. With each hashtag sorted by the most recent posts, I admitted every 2nd post from #LettersAgainstIsolation ($N = 120$), every 25th post from #TheBigSend ($N = 125$) and every 10th post from #Penpalooza ($N = 120$). The resulting corpus comprises 365 original Instagram posts. Table 4.1 provides a breakdown of the letter-writing campaigns studied.

I developed coding categories using a grounded theory approach, which involves 'simultaneous data collection and analysis, with each informing and focusing the other' (Chamaz, 2014: p 508). Among the open coding categories were discussions of digital disengagement, discourses of escapism and nostalgia, writing process versus product, material accumulation, embodiment, labour narratives and audience directives. By iteratively combining and collapsing the codes, I identified key aesthetic themes of these pandemic letter-writing campaigns on Instagram that extended across but also differed between campaign cases (Corbin and Strauss, 2008). These include aesthetic themes of visualizing tangibility, writing to cope and dis/embodied subjectivities. Although all materials in the corpus were publicly available at the time of data collection, analysing them may still have violated users' expectations for how their data will be circulated (Association of Internet Researchers, 2020). I protect users' privacy in this analysis by anonymizing quotations to prevent reverse searchability.

Networked aesthetics of pandemic letter writing

Visualizing tangibility

Among the chief characteristics that distinguish handwritten letters from digital modes of communication is their tangibility. Participants of letter-writing campaigns conveyed the tangibility of their missives by aestheticizing the accumulation of material objects and the labour of craft production on Instagram.

Table 4.1: Letter-writing campaign details

Campaign name and hashtag	Campaign duration	Letter writers	Letter recipients	Communication	Total posts	Sampled posts	Common hashtags
Letters Against Isolation, #Letters AgainstIsolation	April 2020 to present	Volunteers through email list, Google spreadsheets	Senior citizens in care homes	One-way, one-time	241	120	#papercrafting, #cardmaking, #snailmail, #covid19 correspondence, #calligraphy
The Big Send, #TheBigSend	17 October 2020	Volunteers through campaign website	Under-represented voters in swing states	One-way, situational	3123	125	#vote, #vote2020, #letterwriting, #useyourvoice, #bluewave
Penpalooza, #Penpalooza	August 2020 to present	Penpals through Elfster	Penpals through Elfster	Two-way, ongoing	1186	120	#snailmail, #saveusps, #penpal, #happymailswap

Although differences emerged across the campaigns, each of them framed letter writing as a slow, material and intensive mode of communication, distinguishing it from digital connectivity. For example, despite variability in the quantities of letters that participants were encouraged to write and send, each campaign hashtag showcased the accumulation of items associated with letter writing. Users suggested tangibility in their photos by visually and textually indexing processes of production and consumption. At the centre of these processes were the letters and cards themselves. Instagram posts across all three campaigns featured letters and cards as the primary subject matter, including handwritten or typed letters and hand-drawn or store-bought cards.

Photos included evidence of the production process in the form of carefully curated collections of letter-writing supplies as well as letter-writing waste. Participants in all three campaigns photographed piles of letters or envelopes next to collections of pens and paints, stamps and stationary sets. This was accompanied by photos of the by-products of letter writing, such as empty pens and used-up stamp sheets. For example, some Instagram images displayed tables covered with backings from adhesive envelopes. Users also described or quantified the consumption involved in letter-writing practice in their Instagram captions, with one participant in The Big Send recounting '$420 spent on stamps, 53 nights spent writing letters, 10 pads of paper used, and 3 pens that lost their lives along the way'. The quantification of letters parallels wider digital cultures of quantification – such as the quantified self (Lupton, 2016) – which are especially pronounced in Instagram's interface and feedback features (Butkowski et al, 2020). While letter-writing campaigns focus on producing and gathering letters, digitizing them places further emphasis on metrics.

Among the most common images across the three campaigns were completed letters displayed in domestic interiors. People arranged letters, envelopes and supplies in ways that demonstrate both their accumulated quantities and the creative skills that contributed to their production. The letters were often spread out to fill the photo frame with limited visibility of the background, presented adjacent to one another, in stacks, or overlapping. In some cases, participants arranged their letters to display a message, such as the number of letters depicted or the word 'vote'. Letter arrangement was also a privacy-preserving mechanism when letters overlapped and thus obscured recipient and/or return addresses on envelopes. Addresses were also covered with decorative objects, hands or blurred out using photo-editing tools. These methods demonstrate an unspoken recognition that online visibility also begets vulnerability, which can result in unwanted intrusion or exploitation (Duffy and Hund, 2017). In some cases, the campaign Instagram accounts enforced this standard by commenting on photos that were not sufficiently anonymized (with visible addresses) and asking participants to delete and repost the photos with personal information obscured.

Legible text from within the letters themselves was largely absent from the Instagram photos. While letter text was visible in many of the photos, and blurred, obscured or covered in some, it was not the focal point of most images, with the material accumulation of letters taking precedence instead. Images of people reading letters were most common in posts from Penpalooza, which aestheticized 'spending the day reading all of my letters' on lavishly decorated couches or in a cosy reading nook. Ultimately, although they shared their letters using digital photography, participants in these letter-writing campaigns used compositional elements to convey the material tangibility of their mail, which were visibly shaped through their awareness of digital and platformed cultures.

Writing to cope

Campaign participants also described and represented letter writing as a method for coping with anxieties of the pandemic era. The process of creating letters was 'soothing' and 'lovely', 'like free therapy on the days when everything felt so, so sad', according to Penpalooza writers. Emergent aesthetic practices highlighted various strategies for coping with political, social and health-related challenges. Whereas participants in Penpalooza tended toward coping through nostalgia, Letters Against Isolation suggested creativity as an outlet. In contrast, The Big Send cast letter writing as a form of active political participation amid pandemic isolation. Across each of these strategies, letter writing enables a form of emotional release and social connection that participants position as being unsatisfied by digital modes of communication. For example, one Penpalooza participant said:

> 'People all across the globe are connecting in new ways, bridging cultural and geographical space in expressions of love and humanity. It's not all Zoom and digital media though. I love how @penpalooza encourages us all to reach out in the simplest and most personal way: the writing of a letter to a stranger, a fellow human on this planet.'

Just as Menke (2017) argues that people cope with media change by turning to nostalgic media technologies, many letter-writers conveyed nostalgic feelings and imagery in their Instagram posts. Letter-writers expressed positive emotions associated with employing the foregone communication modality of personal correspondence. They included nostalgic imagery in their photos, suggesting time periods when letter writing was a primary mode of communication. Penpalooza conveyed an antique or vintage aesthetic through sepia-toned filters as well as writing supplies such as wax seals and vintage stamps. One Penpalooza participant wrote 'Sending/receiving letters in the fall makes me feel extremely old-timey in the best way', or described the challenge of avoiding their penpal's social media profiles 'so we can get to know each other on paper'. Another imitated dated turns of phrase to suggest a positive future after the pandemic when people can go out again: 'I'm looking forward to saying things like "I'm sorry, but I simply cannot go out tonight. I absolutely must catch up on my correspondence"'.

Others coped with the extra time at home and away from social engagements by pursuing letter writing as a creative project. Instagram posts highlighted the artistic craft of letter writing, especially for Letters Against Isolation. Participants described themselves as 'mail artists', and demonstrated their awareness of wider mail art communities on Instagram through hashtags such as #snailmailideas and #snailmailer. They discussed and depicted creation processes such as collaging found objects, recycling letter-writing supplies, and painting or drawing alone or with children. One Letters Against Isolation participant wrote 'Blew the dust off my drawing pencils to create handmade cards for housebound seniors'. Others explained the effects of the pandemic on their creative process and mental health more specifically. To caption a photo displaying a handmade card for Letters Against Isolation, one person wrote 'For me, this pandemic and everything that has come up with it has been really tough to deal with … I've been trying to use some of my old scrapbooking supplies and materials to make cards and letters and get my mind off it.'

Some letter-writers set out to make tangible changes in their newfound free time, not only by engaging in letter writing as a community service but also as a form of political participation. Largely among writers from The Big Send, mailing letters to potential voters represented taking control of pandemic and US election uncertainties. As one participant framed it, The Big Send enabled them to 'channel my desperation into something contemplative, intimate, and useful'. Another said 'it was a therapy of sorts to write to strangers about how important their voice is 35 times'. This is evinced in the many images of people holding stacks of letters, handing letters directly to postal workers, or placing their letters in mailboxes. In contrast to the other two campaigns, The Big Send centred on people actively mailing letters, demonstrating displays of political agency. These photos coincided with calls to get involved or vote. One participant wrote:

> If you're feeling low about the future, writing postcards and letters as a volunteer to remind people how to vote is a warm place in all the chaos. Just think: there are people checking their mailboxes right now who need to read what your handwriting spells out. You and your vote are valuable.

Letter writing was positioned as a form of political empowerment and therapeutic expression among Big Send participants. More broadly, the engaged disengagement of Instagrammable letter writing enabled the practice and perpetuation of communicative coping mechanisms to weather pandemic stresses.

Dis/embodied subjectivities

Just as letters can augment or stand in for other forms of social connection, they can also stand in for the person who wrote them. Performing digital disengagement can serve as a form of identity construction (Portwood-Stacer, 2012), and displaying letters on social media both indexes personal characteristics and suggests personal subjectivity. Letters, like most other forms of textual communication, are disembodied documents, even if they are also physical and material artifacts. However, according to Milne (2010), the disembodiment of letters lends itself to imagined and potentially more intimate forms of presence and closeness between letter-writers and recipients. Tiidenberg and Whelan (2017) argue that visual self-representation can encompass any image or object that the subject closely identifies with. Because they are the creative and emotional products of campaign participants, sharing letters on Instagram can also serve as a form of self-representation and furnish tools for constructing and understanding the self, even though they do not directly represent participants' bodies or likenesses.

Some people shared selfies in which they hold stacks of letters in one hand and their phone camera (out of frame) in the other. Others shared photos of themselves taken at various parts of the letter-writing process: composing, mailing or consuming letters. However, in some cases the subjectivity and embodiment of the letter-writer is implied rather than explicit (Zappavigna, 2016). Glimpses of disembodied hands covering address information or placing letters into a mailbox suggest the presence of letter-writers even when their faces are obscured. Similarly, the careful arrangement of accumulated letters into stacks and designs suggests the influence of the letter-writer in constructing an image of their work and craftsmanship: 'I put my heart and soul into these letters. It's like sending a part of myself', one Penpalooza participant wrote. Some participants discussed the physical tolls of letter writing in their Instagram captions as well, most commonly in The Big Send. One person said 'My hand hurts, anyone want to give me a hand massage?'. Although participants described letter writing as a therapeutic release, they also sometimes described it as physically exhausting and time-consuming. While participants left their mark through letters and Instagram profiles, their social media posts suggest that the process of letter writing impacted their bodies in ways that digital modes of interaction may not have.

Conclusion

Whether it is self-chosen or externally imposed, digital disengagement is typically categorized as a subtractive phenomenon that involves reduction, breakage and refusal of digital connection. However, it does not necessarily represent a withdrawal from participation in social life. For many people, during a pandemic that required extensive digital media use in educational, professional, political and personal contexts, escape from the digital environments of video conferencing and social media platforms was not possible. This shift, alongside the grief, health concerns, economic hardships, political turmoil and heightened racism that accompanied the pandemic in the US, has had serious implications for mental health and wellbeing on an international scale (Nguyen et al, 2020). For some, resorting to offline and nostalgic media practices, such as letter writing, presented an alternative to solely digital modes of sociality. Letter writing provided a tangible, creative and slow communication outlet and a socially distanced opportunity for community service and political participation. According to the Instagram posts that I analysed, it also served as a 'therapeutic' and expressive coping mechanism for some amid mounting external anxieties and tensions.

In the case of letter-writing campaigns such as Letters Against Isolation, The Big Send and Penpalooza, letter writing provided opportunities to socialize with new people, reaching out to offline connections, and, in some cases, establishing offline social networks. The disengagement of these letter-writing campaigns was inherently partial as all of them originated and were maintained through online systems, such as spreadsheets or email lists. However, documenting, reflecting on and sharing letters on social media platforms such as Instagram enabled people to connect their offline material interactions with their online social networks. In doing so, they also often encouraged these online connections to participate in letter-writing campaigns themselves, both performing their own disengagement and the labour of promoting disengaged practices (Portwood-Stacer, 2012). Ultimately, I argue that, as a practice of partial digital disengagement aestheticized and performed on Instagram, pandemic letter writing demonstrates the additive potential of disengagement from digital interactions. Rather than a choice not to engage in digital sociality, letter-writing campaigns cultivated novel connections that would not have been made outside of epistolary exchanges or pandemic digitalities. Participants then redirected those experiences into their online social networks by aestheticizing their own letter writing.

Despite the additive potential of letter writing as a disengaged practice, it is also restricted to those with digital, monetary and temporal resources. Pandemic over-connectivity was a problem specific to those with internet and device access as well as the luxury to work or learn from home (Treré, 2020). Letter writing at the aesthetic standards typical of Instagram also involves expensive supplies, beyond stamps and envelopes, such as stationery and specialty pens. A major critique of digital disengagement initiatives focused on personal wellbeing and cultural capital is their adherence to neoliberal logics of technology and independence (Kaun and Treré, 2020; Natale and Treré, 2020). As a nostalgic escape from digital over-connection, letter writing endorses these logics rather than opposing or disrupting them. It characterizes disengagement as temporary, even momentary, and situational – a leisure activity for those with the wealth and time to practise letter writing. While performing personalized political acts online, such as supporting the US Postal Service or voting in an election, can encourage participation among one's digital and epistolary networked connections, it also exposes a tension between socially desirable identity construction and impactful political involvement amid pandemic social frameworks. In turn, pandemic letter writing and Instagram sharing present an illustrative case for probing the definitional boundaries of the digital disengagement continuum, particularly within the context of pandemic digitalities (Kunstman and Miyake, 2019). Although this study examines only three of many pandemic letter-writing campaigns, it challenges existing inquiry on digital disengagement, detox, refusal, disconnection and withdrawal to examine particular practices of disengagement and how they may be performed and displayed in the digital realm.

Notes

1. Letters Against Isolation (n.d.). *Letters against isolation*. https://www.lettersagainstisolation.com/
2. The Big Send (n.d.). *The Big Send*. https://votefwd.org/bigsend
3. Penpalooza (n.d.). *Penpalooza*. https://penpalooza.com/

References

Association of Internet Researchers (2020) *Internet Research: Ethical Guidelines 3.0*, Available from: https://aoir.org/reports/ethics3.pdf [Accessed 15 October 2021].

Barton, D. and Hall, N. (2000) *Letter-Writing as a Social Practice*, Amsterdam: John Benjamins.

Basu, T. (2020) 'Letter-writing staved off lockdown loneliness. Now it's getting out the vote', *MIT Technology Review* [online] 18 September. Available from: https://www.technologyreview.com/2020/09/18/1008559/letter-writing-lockdown-loneliness-get-out-the-vote/ [Accessed 15 October 2021].

Bucher, T. and Helmond, A. (2017) 'The affordances of social media platforms', in J. Burgess, A. Marwick and T. Poell (eds) *The SAGE Handbook of Social Media*, Thousand Oaks, CA: Sage, pp 233–253.

Butkowski, C.P., Dixon, T.L., Weeks, K. and Smith, M. (2020) 'Quantifying the feminine self(ie): gender display and social media feedback in young women's Instagram selfies', *New Media & Society*, 22(5): 817–837.

Casemajor, N., Couture, S., Delfin, M., Goerzen, M. and Delfanti, A. (2015) 'Non-participation in digital media: toward a framework of mediated political action', *Media, Culture & Society*, 37(6): 850–866.

Charmaz, K. (2014) *Constructing Grounded Theory*, Thousand Oaks, CA: Sage.

Cooper, S. (2021) 'Penpalooza, the global pen pal project soothing lockdown loneliness', *AnOther* [online] 19 January. Available from: https://www.anothermag.com/design-living/13055/penpalooza-the-global-pen-pal-project-soothing-lockdown-loneliness-rachel-syme, [Accessed 15 October 2021].

Corbin, J. and Strauss, A. (2008) *Basics of Qualitative Research: Techniques and Procedures for Developing Grounded Theory*, Thousand Oaks, CA: Sage.

Danovich, T. (2020) 'Snail mail is getting people through this time', *The New York Times* [online] 1 July. Available from: https://www.nytimes.com/2020/06/24/style/mail-letters-coronavirus.html, [Accessed 15 October 2021].

Duffy, B.E. and Hund, E. (2019) 'Gendered visibility on social media: navigating Instagram's authenticity bind', *International Journal of Communication*, 13: 20.

Gibbs, M., Messe, J., Arnold, M., Nansen, B. and Carter, M. (2015) '#Funeral and Instagram: death, social media, and platform vernacular', *Information, Communication & Society*, 18(3): 255–268.

Jorge, A. (2019) 'Social media, interrupted: users recounting temporary disconnection on Instagram', *Social Media + Society*, 5(4): https://doi.org/10.1177%2F2056305119881691.

Kaun, A. and Treré, E. (2020) 'Repression, resistance and lifestyle: charting (dis)connection and activism in times of accelerated capitalism', *Social Movement Studies*, 19(5–6): 697–715.

Kuntsman, A. and Miyake, E. (2019) 'The paradox and continuum of digital disengagement: denaturalising digital sociality and technological connectivity', *Media, Culture and Society*, 41(6): 901–913.

Leaver, T., Highfield, T. and Abidin, C. (2020) *Instagram: Visual Social Media Cultures*, Cambridge, UK: Polity Press.

Lupton, D. (2016) *The Quantified Self*, Chichester, UK: Wiley.

Menke, M. (2017) 'Seeking comfort in past media: modeling media nostalgia as a way of coping with media change', *International Journal of Communication*, 11: 21.

Milne, E. (2010) *Letters, Postcards, Email: Technologies of Presence*, London: Routledge.

Natale, S. and Treré, E. (2020) 'Vinyl won't save us: reframing disconnection as engagement', *Media, Culture and Society*, 42(4): 626–633.

Nguyen, M.H., Gruber, J., Fuchs, J., Marler, W., Hunsacker, A. and Hargittai, E. (2020) 'Changes in digital communication during the COVID-19 pandemic: implications for digital inequality and future research', Social Media + Society, 6(3): https://doi.org/10.1177%2F2056305120948255.

Niemeyer, K. and Keightley, E. (2020) 'The commodification of time and memory: online communities and dynamics of commercially produced nostalgia', *New Media & Society*, 22(9): 1639–1662.

Portwood-Stacer, L. (2012) 'Media refusal and conspicuous non-consumption: the performative and political dimensions of Facebook abstention' *New Media & Society*, 15(7): 1041–1057.

Tiidenberg, K. and Whelan, A. (2017) 'Sick bunnies and pocket dumps: "not-selfies" and the genre of self-representation', *Popular Communication*, 15(2): 141–153.

Treré, E. (2021) 'Intensification, discovery, and abandonment: unearthing global ecologies of dis/connection in pandemic times', *Convergence*, 27(6): 1663–1677.

Treré, E., Natale, S., Keightley, E. and Punathambekar, A. (2020) 'The limits and boundaries of digital disconnection', *Media, Culture and Society*, 42(4): 605–609.

United States Postal Service (2018) 'A new reality: correspondence mail in the digital age' *United States Postal Service* [online] 5 March. Available from: https://www.uspsoig.gov/sites/default/files/repo rts/2023-01/RARC-WP-18-004.pdf [Accessed 1 April 2023].

United States Postal Service (2021) 'Customer perceptions of the U.S. Postal Service during the COVID-19 pandemic', *United States Postal Service* [online] 9 April. Available from: https://www.usps oig.gov/sites/default/files/reports/2023-01/RISC-WP-21-002. pdf, [Accessed 1 April 2023].

van Dijck, J. (2013) *The Culture of Connectivity: A Critical History of Social Media*, Oxford: Oxford University Press.

Yuan, Y. (2021) '"Think on paper, share online": interrogating the sense of slowness and disconnection in the rise of shouzang in China', in A. Jansson and P.C. Adams (eds) *Disentangling: The Geographies of Digital Disconnection*, Oxford: Oxford University Press, pp 189–224.

Zappavigna, M. (2016) 'Social media photography: construing subjectivity in Instagram images', *Visual Communication*, 15(3): 271–292.

FIVE

Data Minimalism and Digital Disengagement in COVID-19 Hacktivism

Annika Richterich

Introduction

Technological solutionism is appealing, especially during a crisis forcefully asserted to be 'unprecedented' (Milan, 2020; Taylor, 2020; Yan, 2020). The global spread of COVID-19 triggered an avalanche of initiatives promising digital solutions for tackling the pandemic and its impact on societies (Budd, 2020; Madianou, 2020). Typically, the retrieval and analysis of personal, health-related data were/are at the heart of these initiatives (Breuer et al, 2020; Lucivero et al, 2020; Hoffman, 2021). Amidst the flurry of governmental and corporate actions, civic organizations warned that 'efforts to contain the virus must not be used as a cover to usher in a new era of greatly expanded systems of invasive digital surveillance' (Civil Society Statement, 2020). Among those signing the statement were hacker communities such as the Honduran Barracón Digital Hacklab and the German Chaos Computer Club.

This chapter explores activism – aka 'hacktivism' (Jordan, 2002; Jordan and Taylor, 2004; Milan, 2013; Tanczer, 2016; Maxigas, 2017; Galli, 2018; Romagna, 2020) – among such hacker communities, opposing digital solutionism in times of the COVID-19 pandemic and advocating for 'data minimalism' as a means of digital disengagement. Hacktivism, a portmanteau of hacking and activism, refers to the 'politically motivated use of technical expertise' (Milan, 2015: p 550) and has been described as 'activism gone electronic' (Jordan and Taylor, 2004: p 1). I focus on the case of the German Chaos Computer Club (CCC)[1], notably its hacktivism early on during the COVID-19 pandemic (from 2020 to 2021). Conceptually, I relate the notion of digital disengagement (Kuntsman and Miyake, 2019) to data minimalism (Loesing et al, 2010; Vitale et al, 2018; Kaufmann, 2020). In doing so, this chapter emphasizes informed, purposeful strategies for digital disengagement through data minimalism, observed among a hacker association that is known for high levels of digital skills and tech-political expertise.

Especially early on during the pandemic, Germany was among the Western European countries most severely affected by COVID-19. Despite a comparatively well-funded and resilient health system, even in terms of capacities and equipment in intensive care units, hospitals and other healthcare providers were significantly strained due to the pandemic (Rieg et al, 2020). In response, manifold technological initiatives emerged, aiming and claiming to mitigate the impact of COVID-19. Most of these involved collecting and analysing user information, including sensitive data concerning users' health, the places they visit, and the people they encounter, for example. While datafication and its societal implications was already a concern for activists, non-govermental orgaizations and critical researchers before the pandemic, the COVID-19 crisis propelled these concerns to another level (Dencik and Kaun, 2020; Milan et al, 2021; Fan and Fox, 2022). According to Qureshi (2021: p 153), 'the COVID-19 pandemic increased the scale and scope of datafication while reducing the rights of those whose data is harvested ... techno-solutionism, frequent technological interventions, excessive public attention on elaborate yet ineffective procedures and information asymmetries through new applications of technology have made worse existing inequities while revealing the tenuous relationship between authorities and citizens'. Di Salvo (2021: p 164) speaks of a 'datafied pandemic', during which internet and technology access were becoming even more crucial for people's ability to participate in social and economic activities, further worsening the impact of inequalities – not least due to the 'exclusive design of socially-impactful technologies'.

With vaccine developments merely in the works in early 2020, digital technologies were moreover framed as philanthropic, transitory solutions to the pandemic. However, such 'solutions' were often firmly in the hands of private-sector organizations, driven by profit. The digital, data-intensive technologies that emerge in such contexts tend to be rooted in a data expansionist paradigm (Heaphy, 2019; see also van Dijck, 2014). They are founded on and promote the idea that datafication as maximized data retrieval and storage (that is, as unrestricted as possible), is, by definition, preferable and economically as well as societally beneficial (Mayer-Schönberger and Cukier, 2013). In contrast, hacking communities such as the CCC have long emphasized the relevance and benefits of civic rights-preserving technology (Kubitschko 2015a, 2015b), also with regard to COVID-related technology. I argue that 'data minimalism' is central to the CCC's recent hacktivism. Data minimalism, sometimes referred to as 'data austerity' (Kerres, 2020), implies that as little data as needed for a specific purpose may be retrieved, and that such data shall be stored only for as long as strictly necessary. With regard to digital technologies and data in the context of the COVID pandemic, the CCC linked data minimalism to recommendations for 'technological downgrading'. Such considerations regarding technological adequacy refer to the argument that low(er)-tech approaches may be more suitable for certain domains and purposes (see also Lyons, 2014; Coe and Yeung, 2015). In case of the CCC's hacktivism, data minimalism was sometimes considered to be best achieved by avoiding digital 'solutions' entirely, that is, by opting for 'digital disengagement'.

Kuntsman and Miyake (2019) propose that digital disengagement be separated from its frequent conflation with social media non-use and an involuntary or ideologically framed abstinence from digital technology. Likewise, by emphasizing the need for such a 'denaturalization' (Kuntsman and Miyake, 2019), the CCC's hacktivism highlights digital disengagement as an informed choice and matter of agency. The CCC stresses that digital solutions are not per se preferable, but that their adequacy needs to be assessed in relation to specific problems. Based on critical assessments of technology emerging and being proposed in the context of COVID-19, the CCC points out the benefits of data minimalism as informed, purposeful and selective digital disengagement. To elaborate on these arguments, the chapter is further structured as follows. First, I discuss the relationship between digital disengagement and data minimalism in more detail. Linking this to examples of the CCC's hacktivism concerning COVID-19 technology, I then discuss how the hacker association advocates for data minimalism as means of digital disengagement. I argue that the CCC's hacktivism illustrates the relevance of conceptualizing digital disengagement as a matter of agency and a practice of informed choice, rather than considering it as ideologically motivated or as an experience that leaves users unwillingly behind. In conclusion, I reflect on my observations and discuss two main avenues for further work on data minimalism in relation to digital disengagement.

Digital disengagement

When discussing digital disengagement and non-users, we tend to focus on those who are considered to be 'less able' to make use of technology (Macdonald and Clayton, 2013; Leenstra, 2017), such as elderly family members who refuse to use digital devices, or those who are unable to access technology due to geographic and economic divides. Academic research also discusses digital disengagement predominantly in terms of societal exclusion and divides (Helper and Reisdorf, 2017; van Dijk, 2020; Köttl et al, 2021). This is partly related to the urgent issue of digital technology acting as catalyst of age-related exclusion and social injustice, for example. However, digital disengagement may also be a question of agency in the sense of deliberate, strategic and selective/partial non-use (see also Wyatt et al, 2002; Portwood-Stacer, 2013; Light, 2014; Casemajor et al, 2015; Fotopolou, 2016). This aspect appears just as relevant, considering that, among other things, it highlights civic practices and reasons for choosing not to adopt certain technology.

Authors such as Portwood-Stacer (2013), Light (2014) and Casemajor et al (2015) have long argued for an understanding of non-consumption and non-participation as a conscious choice and politically relevant practice. (Social) media refusal and performative non-participation (for example, avoiding any interaction with platforms such as Facebook) can function 'as a form of mediated political action rather than as mere passivity' (Casemajor et al, 2015: p 850). Building on such research, Kuntsman and Miyake (2019: p 902) raised the issue that digital disengagement is 'rarely considered as anything but an aberration, whether temporal, demographic or even ideological'. Observing a common tendency to present digital disengagement in mere contrast to social networking, the authors start from the need to discuss digital disengagement as a matter of agency and social practice. They criticize the fact that current literature, explicitly or implicitly, associates digital connectivity with 'being social' or 'acting socially'. Digital disengagement, in contrast, tends to be presented as abstinence from such social engagement: an abstinence that is often considered involuntary, related to a lack of skills or access, or portrayed as ideologically motivated rather than for practical reasons (Kuntsman and Miyake, 2019: p 904). Such a tendency is problematic in that it potentially buys into the ideology of social media companies, while neglecting the social practices, choices and civic agency underlying digital disengagement.

This chapter highlights recommendations and strategies for selective and informed digital disengagement outlined by the CCC, specifically concerning digital technology that has been (or was supposed to have been) developed with the aim of tackling COVID-19. In this context, suggestions for digital disengagement may be partial and selective, for instance advice to avoid certain kinds of contact-tracing apps, such as those that are GPS-based, or they may be more substantial, highlighting the lack of any need for digital solutions. Reflecting criticism by Kuntsman and Miyake (2019: p 901) against treating digital disengagement merely as an 'ideological aberration', the CCC's recommendations should also not be conceptualized merely as disengagement in an attempt to avoid risks. This is also important, as a focus on risk tends to be instrumentalized to discredit critical civic approaches to technology as overly cautious and inhibiting innovation. The CCC emphasizes that its recommendations are based on risk and benefit considerations. It criticizes a dominant digital-by-default mentality when it comes to approaches to problem solving, which ignores the possibility that low-tech or even non-digital solutions may not only be more secure and safe but also more effective and efficient. Thus, the principle of data minimalism appears to be central to the CCC's reasons for advocating and opting for digital disengagement.

Data minimalism

Data minimalism as a concept has been used to describe approaches and software based on the rationale that data collection should be limited to what is strictly necessary for a clearly defined purpose. In their assessment of *The Onion Router* network, a relay and encryption software for anonymous web browsing, Loesing et al (2010: p 12) define data minimalism as a principle ensuring 'that only the minimum amount of statistical data should be gathered to solve a given problem. The level of detail of measured data should be as small as possible' (see also Vitale et al, 2018; Kaufmann 2020). Data minimalism, as built into software and/or articulated as a guideline, counters data expansionism. Data expansionism may be considered a corollary of 'dataism' (van Dijck, 2014), dominantly underlying recent datafication, digital, data-intensive technology and so-called 'big data' approaches (see also Andrejevic, 2014). It is an ideology that underpins both corporate practices and certain strands of academic research, with dataism showing 'characteristics of a widespread belief in the objective quantification and potential tracking of all kinds of human behavior and sociality through online media technologies' (van Dijck, 2014: p 198). Data expansionist approaches are built on the principle that continuous and untargeted data retrieval and storage – both as unrestricted as technically possible – is a priori preferable and societally beneficial.

In contrast, hacking communities have long promoted ideas of civic rights-preserving technology and data minimalism (Kubitschko, 2015b; Wagenknecht and Korn, 2016; Kaufmann, 2020). Drawing on Lindsay (2003) and Dunbar-Hester (2019), Maxigas (2017: p 845) highlights that hackers' rejection of certain technologies is not only related to nostalgia and efficiency deliberations but also considers how 'the perceived socio-political effects of technology influence its adoption by activist and advocacy groups' (see also Maxigas and Latzko-Toth, 2020). Framing non-use as critical technology choice, the author argues that hackers tend to put less emphasis on the convenience of technology use versus the political implications of such technology, such as how it may affect users' control of their data. Hackers' critical stance towards new digital technology is not merely described as means to keep old technology in place, instead: 'hackers' non-adoption as a high-profile rejection of certain functionalities may lead to the exploration of alternative pathways of technology development' (Maxigas, 2017: p 845; see also Wyatt, 2010).

In this sense, technology rejection also potentially serves as driver for supporting or even developing alternatives. A related, though broader, argument is crucial to the reflections by Bihouix (2020) on use of low(er)-tech options as part of more sustainable ways of living. Questioning ideas of 'green growth' and so-called green technologies, the author advocates for an 'approach of frugality and sufficiency and a cleverer use of technology' (Bihouix, 2020: preface). The author emphasizes that opposing existing, alleged high-tech solutions is key to paving the way for alternatives.

With the onset of the pandemic in 2020, hackers also critically examined technologies promoted as solutions to the COVID-19 crisis, among them the CCC hackers. Drawing on the previously discussed argument by Maxigas (2017), such critical examinations and resistance not only reject particular technologies, but, through their non-adoption, hackers also highlight the requirement for preferable, alternative options and approaches. Countering data expansionist approaches, non-adoption and data minimalism as advocated for by the CCC are related to considerations of technological adequacy and recommendations for 'technological downgrading', opposing a 'digital by default' mindset. The next section briefly introduces the CCC more generally, then analyses its COVID-19-related hacktivism.

The case of the Chaos Computer Club

Formally speaking, the Chaos Computer Club is an association registered in Germany ('eingetragener Verein'). Having been founded in 1981 in Berlin, the CCC is one of the longest established hacking communities and is known for its activist dedication to tech-political matters (Kubitschko, 2015a, 2015b; Wagenknecht and Korn, 2016). The association is organized into 25 regional chapters, with smaller 'Chaos meetings' ('Treffs') taking place in (mostly) cities in Germany, Austria and Switzerland. With more than 5500 members, it draws on decentralized expertise and has established itself as a vocal, hard-hitting critic of tech-political issues in Europe and beyond. The hacker association has long been attentive to issues at the intersection of data and digital technology: for example, being a candid and defiant critic of developments such as the German Patient Data Protection Act (Patientendaten-Schutz-Gesetz) and the recently introduced electronic health card (Elektronische Gesundheitskarte). The community's interventions in matters relevant to data are therefore not specific to the COVID crisis, although the pandemic arguably reinforced hackers' attention on the civic rights implications of new and emerging technology, especially with regard to health-related data.

The arguments and examples outlined in this paper are based on the corpus presented in Appendix 5.1, which includes official statements and related documents published by the CCC between April 2020 and May 2021. These publications have been selectively sampled based on whether they address the CCC's activities and standpoints concerning COVID-19 technology. Drawing on these sources, I argue that data minimalism is key to the CCC's recommendations for digital disengagement in relation to civic rights and data security, particularly in two respects. First, data collection should not by default be considered as a digitally aided procedure, and that data security and safety should be considered as well as the approach's effectiveness and efficiency. Second, if digital data collection turns out to be the preferable choice, the data should be collected selectively in terms of timeframe and the type of information included.

Analysis: data minimalism as digital disengagement

The CCC's hacktivism (Jordan, 2002; Romagna, 2020) combines technical assessments of technology, in the form of solicited and unsolicited audits, with public communication outlining the association's findings and their broader implications. Hacktivism has been described as direct and collective action 'that addresses network infrastructure or exploits the infrastructure's technical and ontological features for political or social change' (Milan, 2015: p 550). Jordan (2002: p 18) links this to politicized practices of hacking, aiming 'to threaten commodification and state control of information'. Having been founded in 1981, the CCC was originally located in what was then West Berlin, an enclave surrounded by communist East Germany and thus in close proximity to a state in which government surveillance was part of citizens' everyday lives. Kubitschko (2015a: p 389) has described the CCC's longstanding hacktivism as 'interlocking arrangements', a combination of 'demonstrating and articulating expertise'. After technically assessing a particular app or software and upon finding vulnerabilities (demonstrating expertise), the association would release a public statement outlining the technical aspects and reasons for the issue (articulating expertise). For example, when assessing the Luca contact-tracing app (also discussed further later), the CCC concluded with regard to craftsmanship flaws and vulnerabilities that Luca key fobs, purchased by the hundreds of thousands for people without smartphones, reveal the complete centrally stored location history with every scan:

> Whoever scans the QR code [using the fob] can not only check in under your name in the future, but also see where you've been checked in so far … The vulnerability [of Lucatrack] is obvious and unnecessary. It demonstrates a fundamental lack of understanding of basic IT security principles. Here – once again – it was knitted with a hot needle instead of building a well-considered solution. (Linus, 2021)

Like the earlier example, most CCC publications include reflections on what the respective vulnerabilities mean for users and how they may be able to protect themselves in the future. Usually, the CCC also contacts the organization involved, requesting a response and for the problem to be solved, if possible, prior to making the issue public. In the past, the CCC has in some cases showcased its technical findings (demonstrated expertise) in more practical and public ways too: for example, as part of the 1984 Btx videotext hack, CCC members found a security flaw that 'allowed them to transfer 135,000 Deutschmark (ca. €68,000) from Hamburg's savings bank to their own donation page. Immediately after the hack, the organization re-transferred the money and reported the incident to the data protection commissioner' (Kubitschko, 2015a: p 391).

Continuing this long tradition of technology assessment from a standpoint of highlighting the need to safeguard civic rights and to protect users' data, the CCC also responded to technology developed in the context of COVID-19. Particular emphasis has been put on the issue of contact-tracing apps (sources 2, 6, 9, 10 in Appendix 5.1) and so-called 'Corona lists' (sources 4 and 5), data collected locally by hospitality venues and stored in cloud systems to support eventual contact tracing. With regards to both, that is contact tracing apps and the aforementioned lists, data minimalism is a key concern. As outlined earlier, on the part of data-collecting entities, data minimalism implies that as little data as needed for a specific purpose is retrieved and that such data are stored only for as long as strictly necessary (Regneri et al, 2019). On the part of users, data minimalism refers to practices aimed at revealing as little information about oneself as possible (Vitale et al, 2018). I argue that the data minimalism suggested by the CCC may be conceptually framed as recommendation for a strategic (partial or more substantive) digital disengagement on the side of users and businesses alike.

For example, after finding vulnerabilities in 'Corona lists', specifically having been able to access allegedly securely cloud-stored data from 87,000 datasets collected by restaurants/cafes, the CCC advised against contact-tracing data being digitally processed by hospitality businesses (see source 5 in Appendix 5.1). Suggesting that we 'think first, digital second', it recommended pen and paper systems for locally collecting and storing data as the most secure option, as opposed to cloud-based systems. This recommendation for digital disengagement amounts to a complete eschewal of digital technology use in the form of cloud-based data storage. It was not only aimed at users, but also directed at hospitality businesses (as business-to-business tech customers) and their approaches to data collection. Here, the emphasis on collecting only the data needed, for as long as strictly necessary, goes hand in hand with an emphasis on considering lower-tech options where possible: not only with an eye to security but also with regard to efficiency and control over future use or otherwise of the data. Rather than starting from the assumption that a digitally supported approach is a priori needed and preferable, the CCC suggested considering the advantages of non-digital systems, especially in terms of data security. At the same time, the association raised the issue of whether a digital 'solution' indeed answers the problem at hand. In a response to the introduction of digital vaccination certificates (see source 1 in Appendix 5.1), for instance, the hackers emphasized that digital certificates are by no means forgery-proof (which is a common argument for favouring these over print certificates). Data minimalist values were also emphasized in this context, stressing that 'a digital vaccination certificate needs to be data-frugal' ('Ein digitaler Impfnachweis muss datensparsam sein'). In this context, 'frugal' refers to a sparing, cautious requestfor user data, requiring only as much information as strictly necessary; interestingly, the term frugality is also used by Bihouix (2020). Here and elsewhere, data minimalism is paired with the suggestion that is should be considered whether a digital approach adds any benefit in comparison with paper-based approaches, for example, that is whether the digital element solves or merely shifts the problem.

Also in response to the popularization of contact-tracing apps, the hacking community highlighted the advantages of data minimalism from a civic rights and data security perspective. The CCC warned that contact tracing as an epidemiological strategy for pandemic control can only be effective under certain conditions and should not be abused as a 'digital promise of salvation' (see source 2 in Appendix 5.1). In its ten requirements for contact-tracing apps, the CCC furthermore demands that 'only minimal data and metadata necessary for the application purpose may be stored. Data that is no longer needed must be deleted' (see source 9 in Appendix 5.1). Notably, the association rejects any technology that would imply a 'creation of central movement or contact profiles', as, for example, with GPS-based contact tracing; likewise, it fundamentally criticizes specific apps, not only due to dubious public–private interdependencies as in the case of the Luca app, but because of insufficient data protection and technical flaws (see source 2 in Appendix 5.1). Thereby, data minimalism is mobilized as a criterion guiding the CCC's assessment of technology, while also substantiating hackers' recommendations for users as well as for companies who are business-to-business customers of technology corporations.

The examples provided highlight a recurring argument by the hacker community: that digital, high-tech solutions are not per se preferable, but that their suitability needs to be assessed in relation to the problem at hand and while considering low(er)-tech options too. Data minimalism as a guideline for data collection in the context of COVID is therefore closely related to an emphasis on considerations for technological adequacy and advantages of 'technological downgrading'. The latter notion refers to the possibility that low(er)-tech approaches rather than high-tech upscaling may be more suitable for certain domains and purposes (see also Coe and Yeung, 2015), notably in being less data-intensive and less prone to security flaws. More generally, this has been influential in shaping deliberations in the information security sector, for example, especially in a post-Snowden era, with print technologies such as typewriters or pen and paper being common tools for preventing leaks and digital espionage (Lyons, 2014). In the case of the CCC's hacktivism, the association highlights that resorting to, or continuing to rely on, non-digital paper-based systems may not only be more secure but also more efficient.

Such recommendations for digital disengagement thus span more fundamental rejections of digital approaches: notably in cases where they may be simply superfluous and merely born out of a digital-by-default mindset. However, they also include strategies for partial digital disengagement, such as avoiding certain contact-tracing apps or means of data collection.

In stressing societal advantages and outlining technical suggestions for data minimalism concerning COVID-19 technology, the CCC opposes the data-expansionist ideology (see also Heaphy, 2019) that goes hand in hand with unquestioned assumptions about a societally beneficial upscaling of digital datafication (van Dijck, 2014). Such assumptions form a subset of technological solutionism, founded on the misleading idea that technological innovation alone is the linchpin for societal problem solving. In advocating for data minimalism and technological downgrading, the CCC thus also envisions a variant of digital disengagement as an informed choice and matter of agency. It proposes digital disengagement through data minimalism, and also by means of technological downgrading. Thereby, the hacker association also confronts technological solutionism as a performative alliance of philanthropy and 'big data' business.

Philanthropically coated tech-solutionism by tech corporations has been notably powerful during the pandemic, with potentially risky technology (often resulting from dubious public–private partnerships and interdependencies) being portrayed as solutions to a global health crisis and its wider societal consequences. The CCC highlighted that, despite the severity of these problems, digital technology should still be scrutinized rather than posited as the most suitable means for tackling pandemic challenges. In questioning what (and how) problems were supposed to be 'technologically fixed', hackers opposed an ideology of the commercially, institutionally and governmentally driven data expansionism that was promoted as a necessary and expedient approach for curbing the pandemic. In rejecting such data expansionism embedded in tech solutionist ideology, the CCC instead advocated for data minimalism as means of rational and informed digital disengagement, as opposed to the latter's common conflation with involuntary or ideologically motivated non-use. In doing so, the CCC also made a critical contribution to public discourse, as it countered issues that authors such as Jasanoff and Kim (2013: p 195) have observed as contributing to the values and ideas associated with technology: a tendency to 'systematically downplay some forms of collective risk-taking, whether economic or physical'. Thereby, the CCC's hacktivism also served as a reminder that the digital technology promoted as solutions by tech corporations, for example, may not be the most suitable method to tackle a problem, and that the alleged solutions are not upfront about related risks concerning data security, civic rights and public–private dependencies.

Conclusion

In this chapter, I have explored how the Chaos Computer Club proposes data minimalism and technological downgrading as means for digital disengagement in response to COVID-related technology. My analysis highlights how digital disengagement may be about an informed (partial) avoidance of certain digital technology and an eschewal of approaches borne out of digital solutionism. This observation further emphasizes that we need to avoid conflating digital disengagement with involuntary non-use (Kuntsman and Miyake, 2019). In stressing the pitfalls of 'digital by default' mindsets (Fotopoulou, 2016), the CCC sheds light on digital disengagement through data minimalism and technological downgrading. Based on technical assessments and risk/benefit analysis, the hacker association provides users and businesses acting as tech customers with a basis for informed digital disengagement. Likewise, it advises caution in too readily accepting the assumption that a digital approach indeed solves the problem it responds to.

The CCC's hacktivism encompasses solicited or unsolicited audits of technology such as contact-tracing apps and public statements concerning its findings. Such statements tend to be sceptical when it comes to digital technology heralded as alleged solutions in the context of COVID-19, criticizing companies and governments for instrumentalizing a 'digital promise of salvation' (see source 2 in Appendix 5.1). Based on their technology assessment of contact-tracing apps, Corona lists and proposals for digital vaccination certificates, for example, the CCC emphasizes that data minimalism should be a fundamental principle of digital approaches and that data minimalism is sometimes best achieved by avoiding digital 'solutions' entirely. Recommendations for an avoidance of digital approaches, for example concerning contact tracing by local hospitality businesses (see sources 4 and 5 in Appendix 5.1), are rooted in an emphasis on technology's implications for civil rights and their efficiency. Similarly, criticism regarding specific data-intensive technology (see sources 2, 6, 7, 8 and 9 in Appendix 5.1) starts from the question of whether the envisioned digital approach does in fact respond to the problem it is meant to solve, or simply shifts the problem while simultaneously creating an array of new privacy related issues. Digital disengagement as an informed choice should therefore be conceptualized – and further explored – as a matter of civic agency and in relation to expertise.

Concluding this chapter, I would like to emphasize two main limitations of this commentary, simultaneously pointing out avenues for further research on digital disengagement. First, while I analysed assessments and recommendations made by the CCC, it is an open question as to what extent such recommendations are indeed reaching the intended audience. It would therefore be relevant to investigate if, where, how and why the proposed digital disengagement is actually being acted upon. Research on this could focus on individual users, but may also address small businesses as customers of technology corporations. Second, I have focused on the Chaos Computer Club as an example, raising the question of to what extent the described recommendations for digital disengagement may be specific to the CCC, given the historical and contemporary context of German (techno-)politics and economics. A comparative perspective, considering hacktivism among other communities (see also Maxigas, 2017), would provide insights regarding to what extent digital disengagement is a broader phenomenon among groups known for comparatively high levels of digital skills. Such comparative insights would potentially provide further insight into means of digital disengagement as an informed choice and matter of agency.

Appendix 5.1: Corpus

Source number	Title	Date, time stamp, author	Source	Link
1	Impfausweise beenden keine Pandemien	2021–05–17 06:30:56, erdgeist	CCC website	https://www.ccc.de/en/updates/2021/impfausweise-beenden-keine-pandemien
2	Luca App: CCC calls for an immediate moratorium	2021–04–13 21:04:42, linus	CCC website	https://www.ccc.de/en/updates/2021/luca-app-ccc-fordert-bundesnotbremse
3	136.000 Corona-Testergebnisse samt persönlicher Daten frei einsehbar	2021–03–18 04:20:23, linus	CCC website	https://www.ccc.de/en/updates/2021/corona-testergebnisse
4	CCC meldet Schwachstellen bei weiterer digitaler 'Corona-Liste'	2020–09–04 06:49:32, linus	CCC website	https://www.ccc.de/en/updates/2020/ccc-meldet-schwachstellen-bei-weiterer-digitaler-corona-liste
5	CCC hackt digitale 'Corona-Listen'	2020–08–28 04:34:28, linus	CCC website	https://www.ccc.de/en/updates/2020/digitale-corona-listen
6	Corona-Tracing-App: Offener Brief an Bundeskanzleramt und Gesundheitsminister	2020–04–24 06:02:45, erdgeist	CCC website	https://www.ccc.de/en/updates/2020/corona-tracing-app-offener-brief-an-bundeskanzleramt-und-gesundheitsminister
7	Offener Brief: Geplante Corona-App ist höchst problematisch	2020–04–24 06:02:45, erdgeist	Linked, open letter signed by the CCC	https://www.ccc.de/system/uploads/300/original/Offener_Brief_Corona_App_BMG.pdf

Source number	Title	Date, time stamp, author	Source	Link
8	CCC analysiert Corona-Datenspende des RKI	2020–04–20 11:57:30, presse	CCC website	https://www.ccc.de/en/updates/2020/abofalle-datenspende
9	10 requirements for the evaluation of 'Contact Tracing' apps	2020–04–06 15:19:28, linus	CCC website	https://www.ccc.de/en/updates/2020/contact-tracing-requirements
10	Joint civil society statement: States use of digital surveillance technologies to fight pandemic must respect human rights	2020–04–03 00:01:03, 46halbe	CCC website and elsewhere	https://www.ccc.de/en/updates/2020/pandemie-menschenrechte
11	Aus der Krise lernen: Digitale Zivilgesellschaft stärken!	2020–04–01 05:21:30, henning	CCC website	https://www.ccc.de/en/updates/2020/zivilgesellschaft

Note: Sources are referenced in the text using the numbers in column 1.

Note

1 Formally-legally, the Chaos Computer Club is an association registered in Germany ('eingetragener Verein'). However, while its central office is in Hamburg, there are 25 regional chapters and numerous local groups ('Chaos-Treffs') not only in Germany but also in Austria and Switzerland.

References

Andrejevic, M. (2014) 'Big data, big questions: the big data divide', *International Journal of Communication*, 8(1): 1673–1689.

Bihouix, P. (2020) *The Age of Low Tech: Towards a Technologically Sustainable Civilization*, Bristol: Bristol University Press.

Breuer, J., Bishop, L. and Kinder-Kurlanda, K. (2020) 'The practical and ethical challenges in acquiring and sharing digital trace data: negotiating public–private partnerships', *New Media & Society*, 22(11): 2058–2080.

Budd, J., Miller, B.S., Manning, E.M., Lampos, V., Zhuang, M. and Edelstein, M. (2020) 'Digital technologies in the public-health response to COVID-19', *Nature Medicine*, 26(8): 1183–1192.

Casemajor, N., Couture, S., Delfin, M., Goerzen, M. and Delfanti, A. (2015) 'Non-participation in digital media: toward a framework of mediated political action', *Media, Culture and Society*, 37(6): 850–866.

Civil Society Statement (2020) 'States use of digital surveillance technologies to fight pandemic must respect human rights', *Human Rights Watch* [online] 2 April. Available from: https://www.hrw.org/news/2020/04/02/joint-civil-society-statement-states-use-of-digital-surveillance-technologies-fight [Accessed: 13 June 2022].

Coe, N.M. and Wai-Chung Yeung, H. (2015) *Global Production Networks: Theorizing Economic Development in an Interconnected World*, Oxford: Oxford University Press.

Dencik, L. and Kaun, A. (2020) 'Datafication and the welfare state', *Global Perspectives*, 1(1): https://doi.org/10.1525/gp.2020.12912.

Di Salvo, P. (2021) 'Solutionism, surveillance, borders and infrastructures in the 'datafied pandemic'', in S. Milan, E. Treré and S. Masiero (eds) *COVID-19 from the Margins. Pandemic Invisibilities, Policies and Resistance in the Datafied Society*, Amsterdam: Institute of Network Cultures, pp 164-170.

Dunbar-Hester, C. (2019) *Hacking Diversity*, Princeton, NJ: Princeton University Press.

Fan, B. and Fox, S.E. (2022) 'Access under duress: pandemic-era lessons on digital participation and datafication in civic engagement', *Proceedings of the ACM on Human–Computer Interaction*, 6(GROUP), 1–22.

Fotopoulou, A. (2016) 'Digital and networked by default? Women's organisations and the social imaginary of networked feminism', *New Media & Society*, 18(6): 989–1005.

Galli, B.J. (2018) 'How ethics impacts hacktivism: a reflection of events', *International Journal of Qualitative Research in Services*, 3(1): 11–20.

Heaphy, L. (2019) 'Data ratcheting and data-driven organisational change in transport', *Big Data & Society*, 6(2): https://doi.org/10.1177/2053951719867359.

Helsper, E.J. and Reisdorf, B.C. (2017) 'The emergence of a 'digital underclass' in Great Britain and Sweden: changing reasons for digital exclusion', *New Media & Society*, 19(8): 1253–1270.

Hoffman, A.S., Jacobs, B, van Gastel B., Schraffenberger, H., Sharon, T. and Pas, B. (2021) 'Towards a seamful ethics of Covid-19 contact tracing apps?', *Ethics and Information Technology*, 23(Suppl 1): 105–115.

Jasanoff, S. and Kim, S.H. (2013) 'Sociotechnical imaginaries and national energy policies', *Science as Culture*, 22(2): 189–196.

Jordan, T. (2002) *Activism!: Direct Action, Hacktivism and the Future of Society*, London: Reaktion Books.

Jordan, T. and Taylor, P. (2004) *Hacktivism and Cyberwars: Rebels with a Cause?*, London: Routledge.

Kaufmann, M. (2020) 'Hacking surveillance', *First Monday*, 25(5): https://doi.org/10.5210/fm.v25i5.10006.

Kerres, M. (2020) 'Against all odds: education in Germany coping with Covid-19', *Postdigital Science and Education*, 2(3): 690–694.

Köttl, H., Gallistl, V., Rohner, R. and Ayalon, L. (2021) '"But at the age of 85? Forget it!": internalized ageism, a barrier to technology use', *Journal of Aging Studies*, 59: https://doi.org/10.1016/j.jaging.2021.100971.

Kubitschko, S. (2015a) 'Hackers' media practices: demonstrating and articulating expertise as interlocking arrangements', *Convergence*, 21(3): 388–402.

Kubitschko, S. (2015b) 'The role of hackers in countering surveillance and promoting democracy', *Media and Communication*, 3(2): 77–87.

Kuntsman, A. and Miyake, E. (2019) 'The paradox and continuum of digital disengagement: denaturalising digital sociality and technological connectivity', *Media, Culture & Society*, 41(6): 901–913.

Lenstra, N. (2017) 'Agency and ageism in the community-based technology support services used by older adults', *First Monday*, 22(8): https://doi.org/10.5210/fm.v22i8.7559.

Light, B. (2014) *Disconnecting with Social Networking Sites*, Basingstoke, UK: Palgrave Macmillan.

Lindsay. C. (2003) 'From the shadows: users as designers, producers, marketers, distributors, and technical support', in N. Oudshoorn and T. Pinch (eds) *How Users Matter: The Co-Construction of Users and* Technology, Cambridge, MA: MIT Press, pp 29–50.

Linus (2021) 'Luca App: CCC calls for an immediate moratorium', *Chaos Computer Club* [online] 13 April. Available from: https://www.ccc.de/en/updates/2021/luca-app-ccc-fordert-bundesnotbremse [Accessed: 13 June 2022].

Loesing, K., Murdoch, S.J. and Dingledine, R. (2010) 'A case study on measuring statistical data in the Tor anonymity network', in I. Eyal and J. Garay *Proceedings of the International Conference on Financial Cryptography and Data Security*, Berlin/Heidelberg: Springer, pp 203–215.

Lucivero, F., Hallowell, N., Johnson, S., Prainsack, B., Samuel, G. and Sharon, T. (2020) 'Covid-19 and contact tracing apps: ethical challenges for a social experiment on a global scale' *Journal of Bioethical Inquiry*, 17(4): 835–839.

Lyons, S. (2014) 'Typewriters, not touchscreens ... security the old-fashioned way', The Conversation [online] 12 November. Available from: https://theconversation.com/typewriters-not-touchscreens-security-the-old-fashioned-way-33846 [Accessed: 13 June 2022].

Macdonald, S.J. and Clayton, J. (2013) 'Back to the future, disability and the digital divide', *Disability & Society*, 28(5): 702–718.

Madianou, M. (2020) 'A second-order disaster? Digital technologies during the COVID-19 pandemic', *Social Media + Society*, 6(3). https://doi.org/10.1177/2056305120948168.

Maxigas, P. (2017) 'Hackers against technology: critique and recuperation in technological cycles', *Social Studies of Science*, 47(6): 841–860.

Maxigas, P. and Latzko-Toth, G. (2020) 'Trusted commons: why 'old' social media matter', *Internet Policy Review*, 9(4): 1–20.

Mayer-Schönberger, V. and Cukier, K. (2013) *Big Data: A Revolution that will Transform how we Live, Work, and Think*, Boston, MA: Houghton Mifflin Harcourt.

Milan, S. (2013) 'WikiLeaks, Anonymous, and the exercise of individuality: protesting in the cloud', in B. Brevini, A. Hintz and P. McCurdy (eds) *Beyond WikiLeaks*, London: Palgrave Macmillan, pp 191–208.

Milan, S. (2015) 'Hacktivism as a radical media practice', in D. Kidd (ed) *Occupy and Social Movement Communication. The Routledge Companion to Alternative and Community Media*, London: Routledge, pp 457–468.

Milan, S. (2020) 'Techno-solutionism and the standard human in the making of the COVID-19 pandemic', *Big Data & Society*, 7(2): https://doi.org/10/1177/2053951720966781.

Milan, S., Treré, E. and Masiero, S. (eds) (2021) *COVID-19 from the Margins: Pandemic Invisibilities, Policies and Resistance in the Datafied Society*, Amsterdam: Institute of Network Cultures.

Portwood-Stacer, L. (2013) 'Media refusal and conspicuous non-consumption: the performative and political dimensions of Facebook abstention', *New Media & Society*, 15(7): 1041–1057.

Qureshi, S. (2021) 'Pandemics within the pandemic: confronting socio-economic inequities in a datafied world', *Information Technology for Development*, 27(2): 151–170.

Regneri, M., Georgi, J.S., Kost, J., Pietsch, N. and Stamm, S. (2019) 'Computing the value of data: towards applied data minimalism', *arXiv preprint arXiv:1907.12404.*

Rieg, S., von Cube, M., Kalbhenn, J., Utzolino, S., Pernice, K. and the COVID UKF Study Group, et al (2020) 'COVID-19 in-hospital mortality and mode of death in a dynamic and non-restricted tertiary care model in Germany', *PLoS One*, 15(11): e0242127.

Romagna, M. (2020) 'Hacktivism: conceptualization, techniques, and historical view', in T.J. Holt and A.M. Bossler (eds) *The Palgrave Handbook of International Cybercrime and Cyberdeviance*, London: Palgrave, pp 743–769.

Tanczer, L.M. (2016) 'Hacktivism and the male-only stereotype', *New Media & Society*, 18(8): 1599–1615.

Taylor, L. (2020) 'There Is an App for That: Technological Solutionism as COVID-19 Policy in the Global North' in E. Aarts et al (eds) *The New Common: How the COVID-19 Pandemic is Transforming Society*, Tilburg: Tilburg University, pp 209-213.

Van Dijck, J. (2014) 'Datafication, dataism and dataveillance: Big Data between scientific paradigm and ideology', *Surveillance & Society*, 12(2): 197–208.

van Dijk, J. (2020) *The Digital Divide*, Cambridge, UK: Polity Press.

Vitale, F., Janzen, I. and McGrenere, J. (2018) 'Hoarding and minimalism: tendencies in digital data preservation' in *Proceedings of the 2018 CHI Conference on Human Factors in Computing Systems*, New York: Association for Computing Machinery, pp 1–12.

Wagenknecht, S. and Korn, M. (2016) 'Hacking as transgressive infrastructuring: mobile phone networks and the German Chaos Computer Club', in *Proceedings of the 19th ACM Conference on Computer-Supported Cooperative Work & Social Computing*, New York: Association for Computing Machinery, pp 1104–1117.

Wyatt, S. (2010) 'Challenging the digital imperative: Inaugural lecture', in K. Bijsterveld (ed) *Science and Technology Studies at Maastricht University: An Anthology of Inaugural Lectures*, Maastricht, The Netherlands: Datawyse, pp 147–174.

Wyatt, S.M., Thomas, G. and Terranova, T. (2002) 'They came, they surfed, they went back to the beach: conceptualizing', in S. Woolgar (ed) *Virtual Society*, Oxford: Oxford University Press, pp 23–4.

Yan, Z. (2020) 'Unprecedented pandemic, unprecedented shift, and unprecedented opportunity', *Human Behavior and Emerging Technologies*, 2(2): 110–112.

SIX

Digital Solutionism Meets Pandemic Imaginaries

Adi Kuntsman

Introduction

This chapter discusses environmental impacts of the
COVID-19 pandemic, and, more specifically, the impact
of the overall shift towards digital technologies during the
pandemic and in post-pandemic times. The chapter builds on
my long-term work on environmental imaginaries of digital
technologies, the idea of digital solutionism in environmental
sustainability, and the paradigmatic myopia with regard to
environmental harms of digital technologies. My discussion
in this chapter is driven by the question. when lockdowns,
social distancing and other safety measures bring accelerated
(and often compulsory) digitization, what space is left – if
any – for environmentally driven digital disengagement, and
what space is left to challenge the violent impact of extractive
digital economies on human and non-human life, when
pandemic safety is imagined predominantly as dependent on
digital devices, connection networks and data?

Digital technologies and the global digital economy inflict
significant damage on both humans and the environment,
and contribute to global environmental injustice, as many
of their harms affect first and foremost the Global South,
whereas digital overconsumption mostly takes place and

benefits those in the Global North. The harms include mining of the rare minerals needed to produce digital devices, the ever-growing amount of ewaste generated by digital devices (many of which are designed to only last a few years), and the energy demands and carbon emissions of internet traffic and data farms as well as of AI, deep machine learning, cryptocurrency and NFTs (non-fungible tokens) (Chen, 2016; Emejulu and McGregor, 2016; Good, 2016; Qiu, 2016; Velkova, 2016; Cubitt, 2017; Maxwell and Miller, 2020; Brevini, 2021; Truby et al, 2022). However, despite a growing number of critical voices, pointing to the environmental and human costs of digital media and communication, the subject remains on the margins of socio-cultural research into digital transformations. In environmental communication, for example, digital technologies are mostly instrumentalized as tools of effective messages, rather than as material entities with their own environmental footprint (but see several notable exceptions: Starosielski and Walker, 2016; Brevini and Murdock, 2017; Shriver-Rice and Vaughan, 2020). In addition, the majority of discussions around environmental sustainability and climate change are shaped by 'digital solutionism' (Kuntsman and Rattle, 2019) – a belief that digital technologies themselves are not only environmentally neutral but also beneficial for environmental protection.

Digital solutionism was already the organizing principle of both academic and public perceptions of the relationships between digital technologies and the environment before COVID-19. Then came the global pandemic, bringing an accelerated adoption of digital technologies in all spheres of social, economic and political life: from the swift introduction of digital contact-tracing systems (later followed by vaccine passport apps) into national public health systems and global control of borders and travel; the sharp rise in video conferencing that came to replace physical meetings and support social distancing; increased media streaming through subscription media giants such as Netflix and social media video-sharing platforms; accelerated platformization of takeaway delivery services; app-ization and automation of public and commercial services previously provided face to face; and expansion of contactless payment systems that has fast-tracked and privileged cashless retail. Implemented rapidly, and often as temporary 'emergency' measures, these developments were then solidified and became the norm in many countries, even when lockdowns were lifted and many interactions returned to being in person. However, while the social and economic and cultural implications of this digital surge – such as growth in the power and profit of tech giants, increased digital surveillance by states, deepening of the digital divide and digital exclusion, to name just a few – have been discussed extensively by both academics and the media, far less attention has been paid to the environmental implications of pandemic digitization. It is this gap that my chapter aims to address.

The main concern driving this chapter is the question of whether the COVID-19 pandemic changed or cemented some of the foundations of digital solutionism in public perception and academic research, and whether any alternative digital environmental imaginaries have emerged since the pandemic began. I have been observing both the emerging academic conversations and the mainstream media discourse around digital technologies, the environment and COVID-19 from the very first days of the COVID-19 pandemic. These observations inform my writing, together with a smaller and more purposefully collected set of data that was prepared specifically for this chapter. The latter is a mixture of focused, small-scale algorithmic mapping of UK media coverage and a pilot systematic review of published academic work using the Scopus database. Both were probing in nature, aiming to take a snapshot of the media and science debates on the one hand, and the burgeoning research on COVID-19 pandemic and digital technologies on the other – to pave the way for larger-scale, and cross-national, future studies.

My analysis focuses on the striking absence of intersections between discussions around pandemic digitization and questions of climate change and environmental degradation. As such, this chapter simultaneously documents the silence around environmental harms of the digital economy in mainstream conversations, and purposefully and deliberately puts forward the few exceptions that do exist – the rare but powerful examples of alternative imaginaries, critical evaluations and calls for reductions of digital dependency, and refusals of compulsory pandemic digitalities. It is here that the paradigm of digital disengagement is central to my approach. The chapter considers two key elements of the digital disengagement paradigm: denaturalizing the digital, and using digital reduction and refusal as a starting point rather than as an afterthought (Kuntsman and Miyake, 2022). This chapter looks at academic and media discourses on the COVID-19 pandemic, digital technologies and the environment. First, I explore whether and how digital environmental harms are mentioned and acknowledged and challenge increased digitalization as both a necessary pandemic measure and as an environmentally sustainable solution. Second, I ask whether digital reduction and digital disengagement are ever considered as a response to digital environmental harms. Conceptually, the chapter foregrounds digital reduction, refusal and opt-out, seeking new visions that promote climate-driven digital disengagement, instead of merely documenting digital harms or ways in which they are ignored.

Digital solutionism and the pandemic

Environmental imaginaries of the COVID-19 pandemic

For the discussion of media coverage in this chapter, I have limited my focus to two 'quality' UK media outlets – the BBC and The Guardian – that tend to regularly publish reports, opinion and analysis pieces about science and the environment (an initial web search showed that local and tabloid UK newspapers did not cover the topic at all). For both the BBC and The Guardian, a DuckDuckGo search was performed using the search terms: 'COVID-19 + climate' and 'COVID-19 and environment', bringing up seven relevant publications in total for each outlet. The search relied on algorithmic outputs by the search engines, and thus may have missed some publications that did not show up in the search results, despite using DuckDuckGo rather than Google search to avoid bias based on Google analytics of my other searches and online activities. Because my aim here is to identify main themes and approaches, an all-encompassing archival search of media databases was less important – having said that, a further study focusing specifically on media coverage of this topic could utilize systematic search tools offered by the BBC and The Guardian's archives. Furthermore, the mapping performed here could be expanded in the future into a larger-scale study covering international media outlets (allowing a more nuanced comparison by country and region), as well as magazines in the fields of media, environmental science, and IT and computing.

When taking a closer look at the selected publications on COVID-19 and climate change/the environment, two topics dominated the discussion. The first one revolved around the question of whether, and to what extent, the pandemic-related lockdowns impacted the environment and what their long-term effect might be. In the first half of 2020, when most countries spent at least several months in quarantine and lockdowns to contain the spread of COVID-19, air and road travel came to standstill. BBC and Guardian articles from that time discussed the impact of lockdowns and remote working on carbon emission and air pollution, as both were reduced dramatically. For example, one of the articles described 'the biggest carbon crash ever recorded': 'No war, no recession, no previous pandemic has had such a dramatic impact on emissions of CO_2 over the past century as COVID-19 has in a few short months' (McGrath, 2020b). Other publications echoed the same sentiment, noting the change to travel patterns during and in the immediate aftermath of lockdowns (Harrabin, 2020; McGrath, 2020a), raising the question of how to make this change permanent, for example, by encouraging cycling and switching to ebikes (Henriques, 2020), and emphasizing the importance of focusing on a greener and more globally just future after the pandemic (Vince, 2020). The second topic was the parallels – or even direct links – between climate change and the rise of a pandemic. Some noted that COVID-19 and climate change are 'part of the same battle' (Frankel, 2020), and that the pandemic is a 'fire drill' for even worse global climate-related crises yet to come (Harvey, 2020). Others noted that the COVID-19 pandemic resulted from the destruction of nature (Carrington, 2020b), and that the pandemic was 'nature sending us a message' (Carrington, 2020a).

In both the BBC and Guardian articles collected, digital technologies were not in the centre of the discussion, nor was their environmental role in climate change explicitly addressed. In some articles, the pandemic reliance on digital technologies during quarantines was not mentioned at all, leaving the readers with the silent assumption that the digitallyheavy quarantine life was, inherently, more sustainable and possibly even carbon-neutral. In another article, digital communication was addressed in passing as a 'holder' of particular behaviours, for example, in relation to social media which became the platform for fake news linking COVID denial and climate denial (Spring, 2021). Another 'holder' may be seen in the mention of online climate activism, whereby digital communication was instrumentalized as simply an inevitable alternative when face-to-face events were not possible: 'The other [change] is quite simply making discussion around climate more difficult as mass events are postponed. Greta Thunberg has urged for digital activism to take the place of physical protests due to the coronavirus outbreak' (Henriques, 2020).

Finally, digital technologies were mentioned as 'data holders'. Two of the articles that discussed the impact of lockdown on carbon emissions made a reference to studies that used new Google and Apple mobility data (Carrington, 2020c; Forster, 2021).

Both the BBC and The Guardian's coverage of COVID-19 and climate change/the environment appear have followed the two main characteristics of digital solutionism with regard to the environment: an instrumental vision of the digital as merely a tool, and blindness to its materiality, its environmental footprint and the uneven global distribution of digital harms. These characteristics were identified and theorized following the systematic analysis of sustainability studies and their relationships to the digital, performed by myself and a colleague several years ago (Kuntsman and Rattle, 2019). In the case of sustainability studies, digital technologies were primarily imagined as effective technologies for monitoring and managing the environment, as tools for education about the environment and for changing people's behaviour, and as powerful repositories of environmental data. In media discussions of the pandemic, digital technologies are very similarly presented as tools and catalysts of specific behaviours, changes in practice, or as holders of useful environmental data. Their own environmental input is never considered at all. Thus, we see the same myopia around environmental harms of the digital itself that we documented in our sustainability studies. For example, while the drop in travel and related carbon emissions repeatedly made the news, the global rise in digital consumptions (and associated rise in electricity use and emissions) was not mentioned at all.

Early days of academic research on digital technologies, the pandemic and the environment

I continued my search for discussions of the pandemic use of digital technologies and how this might impact the environment by looking at the social science publications that emerged in the first two years since the pandemic began. In Summer/Autumn 2021, I performed a pilot systematic search in the Scopus database, using the following search terms:

- COVID-19; digital; climate – giving 59 results of which 12 were relevant
- COVID-19; eco; media – giving 11 results, of which two were relevant
- COVID-19; ewaste – giving only one result
- COVID-19; eco; digital – giving six results, of which one was relevant
- (COVID-19; digital surge) and climate – giving six results, of which three were relevant

The number of results generated was quite small, which isunderstandable for this period, given the duration of the academic peer review and publication process. For that reason, I did not exclude published conference proceedings as a standard systematic review normally would. Relevance was determined by looking at whether COVID-19, digital technologies and environmental/climate concerns were meaningfully discussed. Among the articles found, many were irrelevant, for example, where the keywords were mentioned in different parts of the article but were not discussed in relation to each other, where COVID-19 was mentioned in passing only, or where the use of terms such as 'climate' was metaphorical rather than direct. Once the irrelevant results were excluded, I was left with 19 articles, and, on closer examination, three more were excluded due to lack of relevance. Given the small number of articles found and the very early stage of my research insights, the observations offered here are of a tentative nature. Nevertheless, they serve as a good indicator of what kind of conversations shaped early academic responses to the pandemic, and, more specifically, offer a reflection on the main question that drives this chapter: whether and to what extent has the COVID-19 pandemic changed – or cemented – the foundations of digital solutionism.

Of the 16 articles analysed, two discussed the pandemic and the digital completely separately – one was a special supplement of *The Medical Journal of Australia*, consisting of chapters on health, some of which discussed climate and environmental justice, while another discussed digital transformation (Backholer et al, 2021). The other item, a journal article, curiously pointed to 'climate' and 'digital' as 'meta policies' in the European Union (Grabbe, 2021) and COVID-19 was merely a context in which these meta policies were addressed. Similarly, another discussed the pandemic as merely a context – and a catalyst – of digital acceleration (Kondratieva, 2021).

Eleven of the 16 articles can be placed within the paradigm of digital solutionism. Six of them describe digital technologies as effective tools of environmental management, including modelling, assessment and development of various digital solutions to environmental problems (Poch, 2020; Abramov et al, 2021; Mezzour et al, 2020; Gürdür et al, 2021; Kondratieva, 2021; Shilin et al, 2021). Interestingly, in some of these articles, both COVID-19 and the climate crisis were merely examples of complex crises to which digital technologies were seen as offering an answer. The other five articles discussed the digital as a means of communication, including as a form of adaptive behaviour during the changes brought up by the pandemic (McKinley et al, 2021), an alternative to face-to-face environmental activism (Huish, 2021), a tool for teaching sustainable development during the pandemic and social distancing (Leal Filho et al, 2021), or as a research site where opinions and perceptions about the environment can be studied. The latter category is particularly interesting: one of the articles discusses the rise in climate awareness since COVID-19 and uses Twitter as a tool for analysing public perceptions and as a platform where citizens can be educated to 'new eco-consciousness' (Caldevilla-Domínguez et al, 2021). The other article focuses on the rise of teleworking (working remotely from home) since the pandemic, and analyses Twitter data to understand public perceptions of teleworking and its 'ecological value' (Loia, 2021). Here, again, we see the traits of digital environmental solutionism discussed previously. The lack of discussion of the materiality of the digital itself, and the blindness towards environmental costs of accelerated digitization, are particularly striking in the last two articles mentioned previously, since both focus explicitly on environmental value and social change, but neither social media communication nor teleworking are seem as having a footprint of their own.

Of the entire corpus, only two articles mention such a footprint. One article – albeit seeking no digital reduction or disengagement – notes explicitly that the pandemic has contributed to the rise of ewaste (Rene et al, 2021):

> The rapid changes or upgradation in technologies, IT requirements for working or learning from home during COVID-19, manufacturers releasing new electronic gadgets and devices that serves the consumers comfort and a declension in services has contributed to an increase in the ewaste or waste of electrical and electronic equipment (WEEE) generation ratesextract from the article's abstract.

The other article is a published conference paper that noted the rise in energy consumption due to remote working during the pandemic (and, in particular, the increased reliance on digital technologies), and as a solution, suggested green IT practices, which, as the authors argue, need to accompany motivation of individuals to adopt pro-environmental behaviours (Alla et al, 2020). However, green IT is discussed only briefly, without a detailed analysis of how it would reduce carbon emissions or ewaste and what would be needed to encourage more sustainable behaviours (and whether these behaviours relate to individuals, businesses or developers).

Challenging pandemic digitalities

In the media and academic articles documented earlier, digital solutionism prevailed, with little concern about the environmental impacts of the pandemic's digital surge, or discussions of the global injustice in terms of how digital environmental harms are distributed, and who should be held accountable. In both arenas, the digital was normalized and naturalized – the reliance on digital technologies was not questioned and the need for digital acceleration was not challenged. Of course, part of the reason may be the small size of the sample due to the limited ways in which my searches were conducted. A more expansive systematic search, as well as a comparison of various national media as well as citizen media such as blogs, would be beneficial in a further study. Nevertheless, the findings presented do reflect the broader silence about digital materialities and digital environmental harms that characterizes both public and policy perceptions and academic research. It is therefore particularly important to highlight alternative perspectives and voices. Working within the paradigm of digital disengagement, these voices – even if atypical and exceptional – have the power to denaturalize normative digitalities, within and beyond the context of the pandemic. In this final section, I focus on three such voices: a multidisciplinary study on the environmental footprint of the pandemic, a project on the environmental tolls of media streaming, and a critical intervention into the extractive economy of contact-tracing apps. By bringing these three voices together, my aim is twofold: first, to see whether and how digital reduction and digital refusal are considered and conceptualized, and second, to assess what the articles teach us about a possible conceptual and political shift away from digital normalization to digital disengagement where environmental and human costs are concerned.

Measuring (and reducing) the carbon, water and land footprint of the pandemic's digital surge

In early 2021, Yale News reported (Cummings, 2021) a new study, led by Yale University academics, into the 'hidden environmental footprint' of the pandemic's digital surge. The study, which interestingly did not come up in my Scopus search, generated a publication in *Resources, Conservation and Recycling* journal (Obringer et al, 2021), consisting of a relatively short article, supplemented by a substantial, open access data set of the information collected from multiple countries by the research team. The article noted that the surge in digital activity that had occurred since the early days of the pandemic and was continuing to rise has a significant impact on carbon emissions, water consumption and land usage. Although very short in length and reading more like an editorial commentary, the article is very significant, especially given the overall silence about the environmental impact of pandemic digitization in both research and media, and the overall celebratory approach to digital technologies, which is noted and critiqued. The main argument of the article is simple: despite some environmental benefits observed in the early days of the pandemic such as reduction of carbon emissions due to a halt in travel, increased digitization has its own footprint that needs to be acknowledged, adequately measured, and reduced.

The most important part of the publication is its data set, detailed in a spreadsheet in the Appendix, which offers figures regarding the internet's carbon, water *and* land footprint – an important intervention in the sustainability discourse, which mostly addresses the carbon footprint only. As the authors note 'Despite their environmental significance and contribution to climate change, the water and land footprints of data use have not been well studied' (Obringer et al, 2021). The article offers a way to estimate and calculate these, and includes infographics on the footprint of internet use and how carbon, water and land footprints have increased since the pandemic. Although the article itself is short, its intervention into the solutionist paradigm should not be underestimated. The study of the internet's footprints is comprehensive, covering global footprints as well as those of specific countries: Brazil, China, France, Germany, India, Iran, Italy, Japan, Mexico, Pakistan, Poland, Russia, South Africa, the UK and the US.

The study also measures the increase in these footprints since COVID-19, and breaks it down by application (entertainment streaming such as Netflix, Amazon or Hulu; videoconferencing applications such Zoom and Skype; social media platforms such as Facebook, Instagram, Twitter or Snapchat; as well as various messaging applications and internet surfing).

Perhaps most importantly, the article contains a call for recognition of the digital environmental impact more generally, and, in particular, its acceleration since the pandemic, and provides very clear suggestions as to how these can be mitigated. The lead responsibility, the authors argue, lies with internet service providers who are responsible for limiting the footprint of their products, as well as that of associated data-processing and transmission. At the same time, as companies are unlikely to do that without external pressure, the authors emphasize the importance of policy solutions and public awareness. Finally, the authors explicitly note the need to reduce digital activities, via a range of disengagement practices: turning off, reducing online time, decluttering digital storage, or reducing the size and quality of digital files:

> Society at large should recognize the power of collective action in reducing the environmental footprint of the Internet to avoid paving an irreversible path to an unsustainable digital world. Large-scale adoption of environmentally responsible online behavior by many individuals is vital for combating climate change and promoting sustainability. Making Internet users aware of the costs of online actions and benefits of making small behavioral changes (through information campaigns, behavioral nudges, etc.) is the first step toward promoting sustainable digital behavior. Small actions such as turning off video during a virtual meeting, reducing the quality of streaming services, decreasing gaming time, limiting time on social media, deleting emails and unnecessary content on the cloud-based storage services, or unsubscribing

from email lists can significantly reduce the environmental footprints of Internet use. (Obringer et al, 2021)

Aesthetics of digital reduction and digital sobriety

The Canadian professor of media studies Laura Marks had been researching the carbon footprint of streaming media for a few years before the pandemic. One part of her work involved working across media studies and engineering to understand and assess ICT's role in electricity consumption and the resulting footprint (Marks et al, 2021b). The project, carried out by a cross-disciplinary team, began before the pandemic and was completed just as the pandemic began, showing clearly the substantial impact of ICT, and video streaming in particular, on carbon footprint. Since the start of the pandemic, this footprint has clearly increased. In addition to developing an assessment framework to evaluate the environmental impact of streaming media (which is somewhat similar to the study by Obringer et al (2021), albeit focusing on carbon footprint only), Marks and her team address the complex and contested nature of establishing such an assessment. Instead of merely suggesting tools that would measure the environmental impact of streaming media, they show that there are disagreements regarding what should be measured and what should be included in such measurements, acknowledging the highly politicized nature of any claims made around assessment of digital environmental harms.

When it comes to imagining alternatives, Marks' work makes several important contributions. One of them concerns the technical make-up of media. Marks goes beyond arguing that film and video streaming has an environmental toll that can be reduced by reduced resolution – she develops and promotes an entire aesthetics of 'small-file' eco media. In collaboration with other students and artists, Marks organized a 'small-file media festival' (https://smallfile.ca/ecomedia/) in 2022; although not directly in response to the pandemic, this was beautifully timed in terms of post-pandemic reconsideration of the relationships between digital media and the environment. The festival and its website promote small-file media as a form of environmental aesthetics, and also offer technical advice for compressing videos: 'Why small files? Because streaming media is overheating the planet' (SFMF, 2022).

Beyond their intervention into the practice and aesthetics of digital media creation, Marks and her team have developed an extensive set of 'best practice' guides for managing online footprint (Marks et al, 2021a). Unlike many sustainability initiatives that address either governments and policy makers, service providers or users, Marks' team has developed an entire collection of guides targeting everyone involved – individuals, institutions, media platforms, service providers, data centres, manufacturers and governments – to address carbon footprints, ewaste and disposable design, and user practices. This broad view of digital footprint as a multi-actor problem and a multi-actor responsibility links to Marks' other important intervention, concerning what she calls 'over-engineering' of digital infrastructures (Marks, 2021). The latter intervention, one can argue, truly adopts the spirit of digital disengagement, by showing that the only way to reduce these infrastructures is to reduce digital use, breaking the circle of anticipatory growth. Addressing the debate regarding how much video streaming contributes to the overall footprint of ICT, she notes that the main issue is not the relationship between streaming and other ICT, but the fact that the entire digital infrastructure is built around future large demands and expected market growth:

ICT's infrastructure of networks and data centers was put in place for data-intensive applications like streaming and computation-intensive applications like AI and blockchain. The infrastructure is engineered to *anticipate* future use and spur consumer demand. The argument that streaming only slightly increases electricity consumption naturalizes the notion that infrastructure should be over-engineered. It encourages additional high-data (and computation-heavy) use that will require infrastructure to expand still more. (Marks, 2021)

The more we use digital infrastructures, the more they grow, and thus the only way to reduce digital environmental harm is to reduce the expansion of ICT.

Calling for systemic reduction of ICT design, economy and practice – the argument that resonates with the broader paradigm of de-growth – is also connected to the proposal recently developed by a French think tank, The Shift Project (https://theshiftproject.org) who use the language of 'lean ICT': 'buy the least powerful equipment possible, change them as rarely as possible, and reduce unnecessary energy-intensive uses' (The Shift Project, 2019). The call to make the digital economy 'lean' is linked to what The Shift Project team describe as 'digital sobriety', calling for a move 'from intemperance to sobriety in our relationship with digital technologies' (The Shift Project, 2019). While the general perspective itself is not new – it is also at the heart of decarbonization, energy efficiency and sustainability – what is new here is its application to digital technologies, put forward precisely at the time when the world has become increasingly, excessively, and, one might argue, indulgently digital. Finally, and crucially, the call for digital sobriety is not a self-centred focus of a developed economy: at the heart of The Shift Project's call for lean ICT is a reminder that it is high-income countries alone that bear the responsibility for digital overconsumption.

Against extractive infrastructures

Although clearly aware of the global inequality of digital (over) consumption and the urgency of environmentally committed changes, initiatives such as lean ICT or small-file eco media do not tackle or even attempt to conceptualize the global and racial injustice of digital environmental harms. This injustice is at the heart of the final article discussed here. 'The extractive infrastructures of contact tracing apps' (Aouragh et al, 2020) is a rapid-response short piece written by The Institute for Technology in The Public Interest[1] collective for a special supplement issue of *Journal of Environmental Media* on the COVID-19 pandemic, digital media and the environment. In this article, the authors look at the rise of contact-tracing apps in the early days of the pandemic through the lens of racial capitalism. At the heart of their intervention is a discussion of the extractivist logic of 'big tech' with regard to natural resources, human lives, and never-ending cycles of exploitation and depletion:

> We are specifically referring to the socio-politico-economic practices of extraction as seen in mineral and fossil fuel production, temporary processes often involving landgrabbing. Disguised as 'progress', they allow profits to be made elsewhere and are followed by devastated habitats. They limit the possibilities for livelihoods, resulting in the displacement of indigenous peoples and 'wasted lives' ... [As noted elsewhere] '[e]xtractivism [feeds] itself by ruthlessly exploiting (and depleting) the resources it controls [and] grabbing new resources in order to continue its operation'. (Aouragh et al, 2020: p 9.3)

Their analysis of big tech's extractivism ties together two interventions, discussed earlier: expansion of the carbon footprint discussion into land and water footprints (Obringer et al, 2021), and analysis of the 'over-engineering' and expansion of digital infrastructures (Marks, 2021). Both include important observations regarding the over-encompassing nature of the digital footprint during the pandemic and generally, and the socio-economic logic of digital normativity that cemented pandemic solutionism. However, in focusing merely on the environmental and the infrastructural, their analysis is incomplete in terms of understanding the global racial injustice of digital environmental harms and the ways these disappear from both public and academic discussion. The article by Aouragh et al (2020) focuses on contact-tracing apps, which were presented as a form of 'computational magic' that concealed both the violence of the human/environmental costs of increased phone use that these apps necessitate, and the racialized differentiation of care and protection, which foregrounds some lives while harming and destroying others:

> Contact tracing apps are complicit in global cycles of racial capitalism which are not only manifest in their use but also all the way down to their production lines across different scales of extraction. For sure, the invention and promotion of contact tracing apps will increase the reliance on smart phones which will in turn necessitate an increased extraction of resources and from this increased levels of damage for communities who live with precious metal extraction. These damages include a range of documented environmental and social harms, including human trafficking, exposure to toxins such as mercury, toxic waste impoundments, increased air pollution and destruction of ecosystems. This … this inescapable loop of extraction [has been described] as 'dig, dump, die'. (Aouragh et al, 2020: p 9.5)

Their conclusion takes us through the logic of digital disengagement, not in a sense of how pandemic digitalities can be minimized, contained or otherwise made workable, but instead going right to the heart of compulsory digitality as violence.,They address the urgent need to protect individuals and communities affected by it when they ask: 'What does it mean to develop technologies to combat a virus without considering the larger impact and the extension of extractive economies?' (Aouragh et al, 2020: p 9.5). Their proposed solution goes beyond regulation and best practice, and even beyond the digital itself, into reshaping and freeing lives harmed by extractivism.

To mitigate this harm, they argue 'requires a public discussion which also facilitates that communities can design, test and shape methods for protection not from viruses, but from extractive modes of existence' (Aouragh et al, 2020: p 9.5).

Concluding remarks: mosaic methodologies, digital disengagement and global climate justice

The idea for this chapter grew out of an increased frustration with how the pandemic's digital surge has been viewed as a necessary and positive development. Writing on the subject since the early months of 2020, I kept hoping to see the development of a more sober and critical conversation about pandemic digitalities, but instead observed how issues of environmental concerns and climate justice were absent from most conversations about climate crisis in (post)pandemic times.

Searching for evidence that counters this absence proved difficult, and also revealed a confusion of terminology and a lack of commonly agreed language that can be used to describe the environmental costs of digital technologies. For example, after not finding media discussions of pandemic digitalities when searching for 'COVID-19' and 'environment' or 'climate change', I tried more purposeful searches for 'are digital technologies bad for the environment' and 'digital footprint of COVID-19' on both Google and DuckDuckGo and for 'COVID-19 and digital surge' on Google Scholar. These searches were performed at several points in time, and none brought a significant number of outputs that might indicate that a conversation is taking place. At the same time, the searches did resulted in finds (one or two results per search) that were of direct relevance. For example, when looking for 'digital footprint of COVID-19', I came across an article on 'digital exhaust', describing a growth in digital traces (Leonardi, 2021), which uses the language of environmental pollution ('exhaust') but only as a metaphor (for a similar phenomenon with regard to digital waste, see Savolainen, 2023), an opinion piece on digital footprints and the danger of hacking (Ringel, 2020), and – finally! – a short article from Euronews (2021) on how digital technologies contribute to climate change. These idiosyncratic results may indicate the methodological limitations of systematic reviews of both media and academic research – because the problem at stake does not have an agreed terminology and lacks a commonly acceptable language (or that the language is highly politicized). As mentioned earlier, discussing environmental impacts of digitization is a contested terrain, where the assessment terms used are continuously disputed and challenged, making it a difficult conversation to develop.

I conclude this chapter with some recommendations to develop a way forward. First and foremost, we must begin to think more openly and creatively about the terms used to find relevant research or emerging public conversations, and use multiple search terms and points of entry in the absence of a clear and commonly agreed language to describe digital environmental impacts. Second, we should strive to create a unified language to help bridge siloed conversations that occur separately in engineering and environmental science, media and social sciences, and other areas. Given that different disciplines may use different terms, such as 'footprint', 'harm' or 'impact', and different conceptual frameworks, it is imperative to begin developing an approach that accounts for these differences, without necessarily privileging one over the other, a sort of 'mosaic methodology' that is attentive to different aspects, components, actors, meanings and impacts.

Finally, I argue that any work on digital technologies, the environment and post-pandemic life should be grounded in the paradigm of digital disengagement. This is a paradigm that denaturalizes digitality by questioning the assumption that solutions must always be digital (an assumption that was only cemented since the pandemic). And, most importantly, digital disengagement makes a twofold intervention in the area of forced digitization by identifying who is pushing for and benefits from digital solutions, and by exploring the harms that this digitization brings. It is only through this lens, I believe, that we can begin to re-evaluate and re-imagine the role of digital technologies in (post)pandemic environmental care and global social and climate justice.

Note

1 https://titipi.org/

References

Abramov, V., Lukyanov, S., Korinets, E., Bolshakov, V. and Vekshina, T. (2021) 'Digital tools for seaports geo-information support while climate change and Covid19 pandemic', *E3S Web of Conferences*, 258, 02020.

Alla, K.R., Hassan, Z. and Chen, S.D. (2020) 'The pro-environmental behaviour and the effect of COVID-19 pandemic in Malaysia on green IT practices', in *International Conference on Computational Intelligence*, pp 75–79.

Aouragh, M., Gürses, S., Pritchard, H. and Snelting, F. (2020) 'The extractive infrastructures of contact tracing apps', *Journal of Environmental Media*, 1(Suppl 1): 9.1–9.9.

Backholer, K., Baum, F., Finlay, S.M., Friel, S., Giles-Corti, B. and Jones, A., et al (2021) 'Australia in 2030: what is our path to health for all?', *Medical Journal of Australia*, 214(S8): S5–S40.

Brevini, B. (2021) *Is AI Good for the Planet?*, Cambridge, UK: Polity Press.

Brevini, B. and Murdock, G. (eds) (2017) *Carbon Capitalism and Communication: Confronting Climate Crisis*, Cham: Springer.

Caldevilla-Domínguez, D., Barrientos-Báez, A. and Padilla-Castillo, G. (2021) 'Twitter as a tool for citizen education and sustainable cities after COVID-19', *Sustainability*, 13(6): 3514

Carrington, D. (2020a) 'Coronavirus: 'nature is sending us a message', says UN environment chief', *The Guardian* [online] 25 March. Available from: https://www.theguardian.com/world/2020/mar/25/coronavirus-nature-is-sending-us-a-message-says-un-environment-chief [Accessed 15 September 2022].

Carrington, D. (2020b) 'Pandemics result from destruction of nature, say UN and WHO', *The Guardian* [online] 17 June. Available from: https://www.theguardian.com/world/2020/jun/17/pandemics-destruction-nature-un-who-legislation-trade-green-recovery [Accessed 15 September 2022].

Carrington, D. (2020c) 'Covid-19 lockdown will have 'negligible' impact on climate crisis – study', *The Guardian* [online] 7 August. Available from: https://www.theguardian.com/environment/2020/aug/07/covid-19-lockdown-will-have-negligible-impact-on-climate-crisis-study [Accessed 15 September 2022].

Chen, S. (2016) 'The materialist circuits and the quest for environmental justice in ICT's global expansion', *TripleC: Communication, Capitalism & Critique*, 14(1): https://doi.org/10.31269/triplec.v14i1.695.

Cubitt, S. (2017) *Finite Media: Environmental Implications of Digital Technologies*, Durham, NC: Duke University Press.

Cummings, M. (2021) 'Surge in digital activity has hidden environmental costs', *Yale News* [online] 27 January. Available from: https://news.yale.edu/2021/01/27/surge-digital-activity-has-hidden-environmental-costs [Accessed 15 September 2022].

Emejulu, A. and McGregor, C. (2016) 'Towards a radical digital citizenship in digital education', *Critical Studies in Education*, 60(1): 131–147.

Euronews (2021) 'Digital footprints contributing 'more and more' to climate change', *Euronews* [online] 2 November. Available from: https://www.euronews.com/my-europe/2021/11/02/digital-footprints-contributing-more-and-more-to-climate-change [Accessed 15 September 2022].

Forster, P. (2021) 'Covid-19 paused climate emissions – but they're rising again', BBC [online] 15 March. Available from: https://www.bbc.com/future/article/20210312-covid-19-paused-climate-emissions-but-theyre-rising-again [Accessed 15 September 2022].

Frankel, J. (2020) 'Covid-19 and the climate crisis are part of the same battle', *The Guardian* [online] 2 October. Available from: https://www.theguardian.com/business/2020/oct/02/covid-19-and-the-climate-crisis-are-part-of-the-same-battle [Accessed 18 May 2023].

Good, J.E. (2016) Creating iPhone dreams: annihilating ewaste nightmares', *Canadian Journal of Communication*, 41(4): 589–610.

Grabbe, H. (2021) 'Normative, protective, transformative Europe: digital and climate meta-policies', in C. Damro, E. Heins and D. Scott (eds) *European Futures: Challenges and Crossroads for the European Union of 2050*, London: Routledge, pp 93–108.

Gürdür Broo, D., Lamb, K., Ehwi, R.J., Pärn, E., Koronaki, A., Makri, C. and Zomer, T. (2021) 'Built environment of Britain in 2040: scenarios and strategies', *Sustainable Cities and Society*, 65: 102645.

Harrabin, R. (2020) 'Climate change: could the coronavirus crisis spur a green recovery?', *BBC News* [online] 6 May. Available from: https://www.bbc.co.uk/news/science-environment-52488 134 [Accessed 15 September 2022].

Harvey, F. (2020) 'Covid-19 pandemic is "fire drill" for effects of climate crisis, says UN official', *The Guardian* [online] 15 June. Available from: https://www.theguardian.com/environment/2020/jun/15/covid-19-pandemic-is-fire-drill-for-effects-of-clim ate-crisis-says-un-official [Accessed 15 September 2022].

Henriques, M. (2020) 'Will Covid-19 have a lasting impact on the environment?', BBC [online] 27 March. Available from: https://www.bbc.com/future/article/20200326-covid-19-the-impact-of-coronavirus-on-the-environment [Accessed 15 September 2022].

Huish, R. (2021) 'Global citizenship amid COVID-19: why climate change and a pandemic spell the end of international experiential learning', *Canadian Journal of Development Studies*, 42(4): 441–458.

Kondratieva, N. (2021) 'EU agricultural digitalization decalogue', *Istoriya*, 12: 16–23.

Kuntsman, A. and Miyake, E. (2022) *Paradoxes of Digital Disengagement: in Search of the Opt Out Button*, London: Westminster University Press.

Kuntsman, A. and Rattle, I. (2019) 'Towards a paradigmatic shift in sustainability studies: a systematic review of peer reviewed literature and future agenda setting to consider environmental (un)sustainability of digital communication', *Environmental Communication*, 13(5): 567–581.

Leal Filho, W., Price, E., Wall, T., Shiel, C., Azeiteiro U.M. and Mifsud, M., et al (2021) 'COVID-19: the impact of a global crisis on sustainable development teaching', *Environment, Development and Sustainability*, 23(8): 11257–11278.

Leonardi, P.M. (2021) 'COVID-19 and the new technologies of organizing: digital exhaust, digital footprints, and artificial intelligence in the wake of remote work', *Journal of Management Studies*, 58(1): 249–253.

Loia, F. and Adinolfi, P. (2021) 'Teleworking as an eco-innovation for sustainable development: assessing collective perceptions during COVID-19', *Sustainability*, 13(9): 4823

Marks, L.U. (2021) 'Over-engineered infrastructure anticipates expanded consumption', *Laura U. Marks* [online] 21 October. Available from: http://www.sfu.ca/~lmarks/blog/files/6247e3a44 b97e80a2fdcd94f2281eb6a-14.html [Accessed 15 January 2022].

Marks, L.U., Makonin, S., Przedpełski, R. and Rodriguez-Silva, A. (2021a) 'Best practices', *Streaming Carbon Footprint* [online]. Available from: https://www.sfu.ca/sca/projects---activities/streaming-carbon-footprint/#Bestpractices [Accessed 15 September 2022].

Marks, L.U., Makonin, S., Przedpełski, R. and Rodriguez-Silva, A. (2021b) *Tackling the Carbon Footprint of Streaming Media*, Ottawa, Canada: Social Sciences and Humanities Research Council. Available from: https://www.sshrc-crsh.gc.ca/society-societe/community-communite/ifca-iac/evidence_briefs-donnees_probantes/earth_ca rrying_capacity-capacite_limite_terre/marks_makonin_przedpelsk i_rodriguez-silva-eng.aspx [Accessed 15 September 2022].

Maxwell, R. and Miller, T. (2020) *How Green Is Your Smartphone?*, Cambridge, UK: Polity Press.

McGrath, M. (2020a) 'Coronavirus: air pollution and CO_2 fall rapidly as virus spreads', *BBC News* [online] 19 March. Available from: https://www.bbc.co.uk/news/science-environment-51944 780 [Accessed 15 September 2022].

McGrath, M. (2020b) 'Climate change and coronavirus: five charts about the biggest carbon crash, *BBC News* [online] 6 May. Available from: https://www.bbc.co.uk/news/science-environm ent-52485712 [Accessed 18 May 2023].

McKinley, E., Crowe, P.R., Stori, F., Ballinger, R., Brew, T.C. and Blacklaw-Jones, L., et al (2021) "Going digital' – lessons for future coastal community engagement and climate change adaptation', *Ocean and Coastal Management*, 208: 105629.

Mezzour, G., Mannane, K., Benhadou, S., Benhadou, M. and Medromi, H. (2020) 'Digitalization against the new outbreak', World Conference on Smart Trends in Systems, Security and Sustainability, London IEEE, pp 707–714.

Obringer, R., Rachunok. B., Maia-Silva. D., Arbabzadeh, M., Nateghi, R. and Madani, K. (2021) 'The overlooked environmental footprint of increasing Internet use', *Resources, Conservation and Recycling*, 167: 105389.

Poch, M., Garrido-Baserba, M., Corominas, L., Perelló-Moragues, A., Monclús, H. and Cermerón-Romero, M., et al (2020) 'When the fourth water and digital revolution encountered COVID-19', *Science of the Total Environment*, 744: 140980.

Qiu, J.L. (2016) *Goodbye iSlave: A Manifesto for Digital Abolition*, Champaign, IL: University of Illinois Press.

Rene, E.R., Sethurajan, M., Kumar Ponnusamy, V., Kumar, G., Bao Dung, T.N., Brindhadevi, K. and Pugazhendhi, A. (2021) 'Electronic waste generation, recycling and resource recovery: technological perspectives and trends', *Journal of Hazardous Materials*, 416: 125664

Ringel, C. (2020) 'Why it's important to be mindful of digital footprints during the COVID-19 pandemic', *The International Association of Privacy Professionals* [online] 29 May. Available from: https://iapp.org/news/a/why-its-important-to-be-mind ful-of-digital-footprints-during-the-covid-19-pandemic/ [Accessed 15 September 2022].

Savolainen, L. (2023) 'Dirty, toxic, dumped: waste as data metaphor', in A. Kuntsman and L. Xin (eds) *Digital Politics, Digital Histories, Digital Futures*, Bingley, UK: Emerald.

Shilin, M., Sikarev, I., Baikov, E., Gogoberidze, G. and Petrieva, O. (2021) 'Digitalization wildfire management near smart city', E3S Web of Conferences, 258: 01003.

Shriver-Rice, M. and Vaughan, H. (2020) 'What is environmental media studies', *Journal of Environmental Media*, 1(1): 3–13.

Small File Media Festival (2022) August 9-14, Simon Fraser University. Available from: https://smallfile.ca/ecomedia/

Spring, M. (2021) 'Covid denial to climate denial: how conspiracists are shifting focus', *BBC News* [online] 16 November. Available from: https://www.bbc.co.uk/news/blogs-trending-59255165 [Accessed 15 September 2022].

Starosielski, N. and Walker, J. (eds) (2016) *Sustainable Media: Critical Approaches to Media and Environment*, New York: Routledge.

The Shift Project (2019) '"Lean ICT: towards digital sobriety": our new report on the environmental impact of ICT', *The Shift Project* [online] 6 March. Available from: https://theshiftproject.org/en/article/lean-ict-our-new-report/ [Accessed 15 September 2022].

Truby, J., Brown, R.D., Dahdal, A. and Ibrahim, I. (2022) 'Blockchain, climate damage, and death: policy interventions to reduce the carbon emissions, mortality, and net-zero implications of non-fungible tokens and Bitcoin', *Energy Research and Social Science*, 88: 102499.

Velkova, J. (2016) 'Data that warms: waste heat, infrastructural convergence and the computation traffic commodity'. *Big Data & Society*, 3(2). https://doi.org/10.1177/2053951716684144.

Vince, G. (2020) 'After the Covid-19 crisis, will we get a greener world?', *The Guardian* [online] 17 May. Available from: https://www.theguardian.com/environment/2020/may/17/after-the-covid-19-crisis-will-we-get-a-greener-world [Accessed 15 September 2022].

SEVEN

State Violence, Digital Harms and the COVID-19 Pandemic: Imagining Refusal, Resistance and Community Self-Defence

Seeta Peña Gangadharan and Patrick Williams in conversation with Adi Kuntsman, Sam Martin and Esperanza Miyake

Seeta Peña Gangadharan is a media and communication scholar and one of the co-founders of Our Data Bodies (https://www.odbproject.org/), a US-based team exploring the ways that digital information on low-income and racialized communities is collected, stored and used by governments and corporations. She has written extensively on the relationships between technologies, and exclusion, marginalization and injustice.

Patrick Williams is a UK-based critical criminologist who has focused, for many years, on racial discrimination and racism shaping the Criminal Justice system. More specifically, his work addresses the over-policing and over-criminalization of Black, Asian and other minority ethnic communities in the UK. Most recently, he has published a number of articles and reports on 'data driven policing' and on encroachment of tech into structures of institutional and state violence.

Adi Kuntsman: The three of us are very familiar with your work and are much inspired by it. We want to open this conversation by asking you generally about your work on the relationships between state violence, such as policing or warfare, and systemic injustices of racism, and enforced uses of digital technologies. How do these feature in your work, and how do you see the possibilities, or maybe lack thereof, for escaping or refusing the digital and what that might depend on?

Patrick Williams: First, a quick overview of some of the work that I have been involved with, and just as a way of making sense of some of this, it's important for me to acknowledge that the analysis of tech was never really at the fore of the work I was involved with – which is why I am quite pleased to be in this space with Seeta who has a much better understanding of these issues than I do. My work, for 20 years, has been concerned with the over-policing, racial disparities, discrimination and racist police in the criminal justice and law enforcement systems. However, the more recent pieces of work that I have been involved in have forced me to begin to think about the emergence and encroachment of technology into policing. While we acknowledge that there has always been racial disproportionality within the criminal justice systems – and what I mean by that is that racially minoritized groups are more likely to encounter the police, more likely to be pulled into criminal justice and law enforcement systems, more likely to be held in custody and more likely to be subjected to punitive punishments. In recognizing that, there has been a sense of beginning to make sense of how technology, may contribute to that disproportionality, that disparity. In my report 'Being Matrixed' [Williams, 2018], what we found was that local police forces and the Metropolitan Police in London specifically had begun to build databases of individuals who are thought to be 'nominals' or gang suspects, at risk of gang involvement, or are associated with people involved in gangs (in London, the database was named the 'Violent Gang Matrix' – Matrix for short). What we found striking was that there was an over-representation of racialized minority people, representing 90 per cent of people in the database.

If we think about the drivers of criminality and violence, what we can see is that similar groups or racialized minority groups are not particularly involved in violent crime. So, we were concerned with what was taking place. In addition to that, there was a sense that individuals' details were being captured and freely shared across criminal justice agencies. And what was most pertinent for us was that individuals whose details were recorded were not necessarily made aware of their details being held on this database. So, in essence, we began speaking to individuals who have this perception of 'being matrixed' – individuals who have this perception that their details have been captured, and that they were being policed that way. And what we found was that individuals spoke to this notion of being subject to higher levels of policing. For example, one individual spoke about being stopped and searched 200, 300 times; individuals spoke about been excluded from schools; one individual spoke about family members being removed from the family home. Another individual spoke about teachers seeing them at college on one occasion, and then the headteacher calling them in and saying that the individual is going to have to leave because they 'pose a risk to other students' within the college.

So, we found that, despite individuals being unaware of their matrix status, they began to sense this notion of surveillance and regulation as a consequence of being matrixed – being recorded onto this database. And that was beginning to harm their abilities, the ways in which they could navigate their environments, their communities, access resources, etc. And that became the standpoint for us – and the question therefore becomes, how do we make sense of the use of data, and, in particular, we were interested in the fusion of data, organization's data that comes from a range of different sources, and how that may be utilized and manipulated by the police, criminal justice systems and the state generally, as a way of regulating racial minority groups and individuals.

Seeta Peña Gangadharan: I guess I came to the issues about the intersection of technology and justice by, in some ways, a very familiar path for those of us who believe in alternative and independent media and the power of communication and the capacity for social movements to activate and transform different ways of thinking in the general public. And so I came from, in the early late-90s and early-2000s, really thinking about the potential of tools of communication to assist movements for social change, as we called it back then. Right now, we talk very explicitly about racial, social and economic justice. I spent many years trying to advocate, and also I did research on how we can get tools of communication into the hands of different groups that have been shut out of mainstream narratives. This includes young Black men in the San Francisco Bay Area who are trying to get their stories out and not be perceived as 'super predators', as the term was used in the 90s, to other kinds of causes, such as thinking about anti-corporate globalization and reproductive rights, and things of that nature, so really thinking about getting the tools and helping to shape the communicative practices of these groups that were really otherwise shut out and shut down.

By the first term of the Obama Administration, there was a lot of enthusiasm about building out communication infrastructures in the United States; a big cash injection for building broadband tools. And it was just at the time where (mostly libertarian) groups of activists and legal scholars were starting to talk and complain about lack of privacy. There was a lot of attention to privacy, but somehow completely devoid of historical context for anything palpable to people's daily experiences: not only with respect to privacy, but with respect to technology, and how that interacts with people's lives.

It is from there that I started really thinking long and hard – and others were as well – 'Oh, no! What happens when we gain access and are incorporated into this vast communications infrastructure, and those tools aren't necessarily the best tools for liberation and for community self-reliance and for collective self-determination?' And I really started to, again with others, think about this and worry about this; and in the work that I co-organized and co-founded with 'Our Data Bodies' [https://www.odbproject.org/] – which was a pretty massive participatory collaborative research project – it was there and then that this intersection, in terms of being plugged in, always on, being invited to bridge the digital divide and being digitally included, came with this incredible contradiction. Because these are coercive tools. These are tools of hypersurveillance. These are tools of over-policing. These are tools of carceral technologies, as Ruha Benjamin calls them [Benjamin, 2019].

You cannot ignore that in the work that we were doing, and in the work that we continue to do. Whether it's returning and re-entering into society, people that we talked with having this trepidation about information that's being collected by corrections officers, and then welfare – social workers and employers – taking, taking and taking and never reciprocating. There is vast asymmetry that becomes apparent as you are re-entering, that becomes apparent as you're navigating through a system to get a bed to sleep in; not just a temporary bed, but more permanent housing. This asymmetry that appears when you are dealing with the shutdown of basic utilities, or the threat of foreclosure such as the municipal authorities coming to take your house, or the police profiling you because of your insurance that you didn't pay because you have something else that is more immediate and more necessary.

So, these are the kinds of experiences that have been part of the work that Our Data Bodies has done, and I think has made this struggle, where social injustice, and racial and economic injustice kind of intersect with the injustices caused and conditioned by an inherently coercive communications and technological infrastructure. You could not ignore it.

In terms of the second part of your question, about refusal. We were talking about it the other day in one of our team meetings for Our Data Bodies, thinking about refusal and thinking about resistance. One of the things that Mariella Saba, one of the other co-founders, was saying, was that there is refusal that you can do in the moment. You can give a police officer an alternative spelling of your name and they might not find you in the police database. Or you can become familiar with the coordinated entry system in Los Angeles that determines your 'risk score' and what kinds of social services you need. You can learn from the kinds of questions that you are getting – what makes a high risk score according to them – in order to get the services you need. So, there are kinds of refusal and workarounds that you can do in the moment. And then I think often, where we talk about these larger attempts at resistance – an intergenerational vision of what it means to challenge and overcome and replace the institutions that have tried to coerce, control, punish, degrade and dehumanize marginalized people for many generations. So that is where our thinking is at, and we also recognize that this is a long-term struggle. Thinking about resistance and refusal – you always have to anticipate what's going to change and how things are going to be different. So, it is possible, and it is challenging.

Esperanza Miyake: That is really brilliant, both of you. There is an overlap, even though from different perspectives. I would like to ask you now how the COVID-19 pandemic has changed, intensified or altered some of these issues that you both raised. For example, Patrick, you said about being matrixed and heavy policing … During the COVID-19 pandemic, there was also a certain 'health matrix', if you like, with having to collect data from people, such as contact tracing, and how that could potentially be mis-used after the pandemic. Or even the pandemic itself. COVID-19 was quite racialized, in the UK for example, with fines for lockdown violations which were highly racialized, so was policing. So,

I wonder whether there was a sort of overlap there. And Seeta, I was thinking also about how in the media there were a lot of debates about COVID-19 as the great equalizer – because everyone had the same technologies, which meant that there was more access for marginalized groups. Contradicting that, you mentioned that, with regard to the digital tools, there was a sense that it was liberating but equally it was also about surveillance. And you mentioned the possibilities of refusal. In an age where we are being asked for data to contain the virus and to reduce deaths, how much space there is for such refusal or flexibility of refusal?

Seeta Peña Gangadharan: There is so much in that question. Let me see if I can think of a story that will thread through the different pieces. I guess, with COVID and this discourse around equalizing, in terms of the groups and the communities and the people that are part of Our Data Bodies, that was just a false claim from the beginning. I remember my fellow co-founder Tawana Petty has spoken extensively about this in the context of Detroit where she was witnessing both this severe collapse of the public health system in Detroit, alongside the increase or what we can call 'surveillance mission creep' … the use of real-time facial recognition cameras in Detroit to enforce the lockdown restrictions. This idea that the communications infrastructure could be taken over and used to ends that weren't originally part of the equation – we were seeing this unfold in Detroit. We were seeing this unfold in other cities throughout the United States, where okay, we can put a different strategy around these technologies in ways that are debilitating towards people's freedoms and sense of dignity.

This was never about equalizing. And I guess we only have to look at the multimillion-dollar profit in this crisis moment – we as a society became way more dependent on those companies than ever before. They referred to it as 'blitz scaling', where you ramp up because you know there's a 'network effect' and everybody is, like, 'We love this product'. During the pandemic, and in the height of the lockdowns, everything went digital. Everything went to the cloud. Everybody is on Zoom or an equivalent platform. And that isn't only about the data. That is about the very infrastructure that theseompanyies have created and maintained and now deem as 'essential' to the basic functions of democratic societies. Not even just democratic societies, I mean we've seen this in anti-democratic societies, but actually they've implemented a course of communications infrastructure way before that.

What I think is starting to become apparent in that configuration is that now, the stories that we heard on the ground, about how do you protest this stuff, such as this decision that you made about me which came through some automated calculation – that's wrong. So now that things are even more digital, they're more automated. It's not only something that you have to complain about at the social security office, that something has gone wrong and you're being denied certain services. It's not only that you have to deal with the usual bureaucracy of public institutions. You have to now also deal with the opacity of these tech companies and the opacity that they purposefully create because it allows them to control much more.

With respect to refusal, this poses an incredibly enormous challenge, partly because we're a team of people who sit in different communities of practice. I think we're starting to try and draw these threads, where I see problems in computational infrastructure that are growing and growing. It is not just about data but also the configuration of power that exists in part because of companies like Microsoft, Amazon and Alphabet, and they, in part, give oxygen to the other companies that then create facial recognition tools, matching tools, matching databases that are done in real time to [use in] predictive policing tools. What have I learned by being adjacent to these computer scientists, but also what I've learned from the work of the 'Stop LAPD Spying Coalition' – which is part of Our Data Bodies – is that these ecosystems are vast, they are complex, but you can find them. Drawing that thread is really, really important, because it's not as if you just attack the academics that are working on predictive policing tools, and you criticize all of this that's happening, and then it gets unravelled. You have to attack it at different parts. You have to find the pain points that exist within this ecosystem. And that requires a lot of coordination. That requires a lot of time and dedication.

Patrick Williams: That was really valuable for me hearing you speak to that, Seeta. I remember speaking to some of the techies about the consequence of coronavirus – in fact, it was Eric Kind who said to me that the COVID-19 pandemic is possibly to the biotech industry as what 9/11 was to the military–industrial complex. There is something really interesting in terms of what was happening there, and just listening to you speak almost made sense of that oxygen that's been created in this space, in which there is a 'normal-ness' – really quickly, a normal-ness – to these tools. Here is a short anecdote. I remember going out for something to eat with the family and the restrictions were being reduced, but we needed an app to get into the restaurant to get a drink. My daughter's boyfriend said his phone had died, so we couldn't get on to access the app, and that was the basis upon which we couldn't go into the restaurant, or he couldn't. And, in that moment, I thought about those individuals who may not have the phone to use the app, and who are now being locked out and being excluded. And it's always a conversation I've tended to have around the tech, because for me, it's about who is it designed to exclude? And that forces me to begin to think about those groups and communities who have always been perceived with suspicion, always been viewed as the outsider, and the extent to which the risk-making features of tech, but also who is to be excluded, come to the fore. I think it's deliberately designed to exclude my daughter's boyfriend from coming into the restaurant: young Black male will be scored on his sneakers, and in essence, it just made sense. Now, rather than say a doorman (or security guard) knocking you back, the tech is now saying we can't let you in – and there's something disconcerting there for me.

But there's something wider – we did some work on policing during coronavirus, and it forces us to begin to think about those experiences of being policed. And what again we found was – and this is really interesting for me as it's the notion of continuity – what we heard very quickly was that there was an increase in the numbers of people who were being subjected to stop searches; an increase in those who were subjected to financial penalties; there was an increase to approximately 13 per cent in the use of violence and the use of force during coronavirus. And there was something about how individuals made sense of that. And for me, something I want to foreground on the table as well because I was reflecting on this earlier today, there's something about the coronavirus, or the pandemic, almost seeding the idea that we all potentially pose a risk, that everybody potentially was suspicious. It's probably the first time I walked down the road and walked with a wide berth around an elderly white couple, because everyone had this sense of the potential risk of coronavirus and contamination very early on. Something about the seeding of risk and suspicion in all of us, which almost lends itself to 'well, there's an app for that' and 'there's a way in which we can manage that, and there's a way in which we can respond to that'. And we can exclude those who are risky by ensuring that people downloaded two free NHS apps – I don't know how many there are now – as a way of navigating society. So rather than what coronavirus has given us in terms of the tech etc is that sense of seeding this notion that we will all be susceptible to this notion of being risky, and suspicious, and therefore we need to arm ourselves with the tech as a way of being able to navigate society. What we also found was that notions of being policed have changed, shifted beyond the notion of stop and search on the street, and again, speaking to some of the things that we raised earlier, there are individuals who have been out on protest, who were conscious of the police driving up and down the street past their homes, and that sense of being surveilled from the police van. So, the response to the Black Lives Matter protests that were taking place, many individuals had this very different sense being policed in that moment.

For me, it feeds into that idea of legitimizing suspicion, these stop and search and police encounters, so individuals could be stopped and made to account for their presence, within this moment, and it was almost a sense that that was legitimate, it was suspicion-less, but again individuals found a way of making sense of it; grappling with it:

> Why have you stopped me? – You smell of cannabis – But I don't smoke cannabis.
> Why have you stopped me? – We are in tier 4 lockdown – But I am going to work.

This sense of individuals trying to make sense of the encounter in this moment, which is very different. So those traditional arguments that could have been put forward, that the police are discriminating and stopping and searching all young Black people, almost gave way to the sense of almost beginning to question ourselves, well why is this taking place, and why is that stop taking place in that moment. And I think it's therefore the increase in a shift towards the suspicion-less stopping searches which I think is a concern, and also that sense of riskiness being attributed to lots of individuals.

What's also interesting there – and I am thinking about this notion of refusal – is that individuals who therefore did not engage, or individuals who didn't comply with the stop and search, noticed very quickly that there was an escalation in that encounter, a very quick escalation. And we are not sure whether that's the consequence of police themselves being fearful of the risk that was being promoted around society, but the encounter very quickly escalated. For the individual themselves to attempt to try and de-escalate the counter because of that fear of criminalization and violence, that was taking place in those moments with these individuals.

For me from the coronavirus moment, I think it's a shift towards almost a consolidation in intensification of policing. That sense of riskiness, that sense of suspicion and that seeding of those ideas within individuals, and I mean all individuals, has now lent itself, and found itself within the Police, Crime and Sentencing and Court Bill, where there is an explicit notion of stop and search. I think these ideas are now also captured within these requirements for all public-sector bodies to share data with the police, which again is a feature of the Bill. So rather than how the tech is paid for, I think from my position is that sense of how the circumstances have come together to legitimately hold this data and use this data towards predictive policing. The consolidation of racialized tropes of who is responsible for crime. For example, sections 2 and 10 of the Bill speak about racialized constructions of violent crime and violent individuals. It speaks about 'knife crime prevention' orders; ideas of foresight. So the police can legitimately and pre-emptively go and stop and search individuals who may have previously been involved with certain behaviours. It is an intensification/consolidation of these police powers, which I am finding really problematic.

For me, what the pandemic has facilitated is a sense of well, we now have this tech, this ability to use this technology, these skills as a way of reducing the riskiness within society or responding to those ongoing concerns that members of society may have. So we begin to see the entrenchment of these approaches going forward. While I wouldn't hinge the conversation around the pandemic itself – because we're seeing the continuity in police brutality and the use of databases and tech anyway – what we're seeing is that the legislation is almost concretizing these ideas, and almost setting them up for the future. And again, just to echo the point that was made earlier about the big tech companies that are seeing this as an opportunity to cement their position and the seductions of technology, I don't see this disappearing very quickly. I suppose that would be the point of the next question, won't it?

Sam Martin: Yes, absolutely, you have pretty much solidified what we've been thinking as well of this intensification of surveillance, data sharing and policing using these notions of 'risk'. We're wondering, what is the possibility for digital refusal, and trying to fight for social justice, now that this web, this matrix of health and big data, has been woven so tightly, and, with politicians trying to legislate this, and with the big tech companies saying it's absolutely necessary that you have, for example, a vaccine passport to travel to go to a local event. If people are refusing and saying 'Well, actually, I can't get a vaccine for health reasons' or 'I'm worried about giving my data and sharing my data with a random restaurant that I need to go into – how can I trust, what you're going to do with my data versus what the government is going to do?' So, with this in mind, what do you think are the possibilities for people to refuse in the moment? As you were saying, Seeta, and also looking forward in terms of resisting what's going on – if everything feels like it is closing in relying on COVID as this backdrop – what kind of ways, do you think people could fight against this, if possible?

Seeta Peña Gangadharan: I have many ideas. Some of them are probably totally wacky and un-implementable. But I have a couple thoughts on this, and your question actually brought me back to thinking about that, so what was it like when passports were introduced originally, and borders were introduced. Because those aren't things that existed before, if you lived in the Ottoman Empire, you were as likely to go to Istanbul as you were to Athens, or somewhere else in Greece. There was fluidity. And so yes, you've got me inspired to think historically about what was going on in those times to push back against the consolidation of land or borders, the consolidation of how we organize and control mobility in society. But in terms of when you have collusion between the state and private power, as vast as it is today, it feels grim.

If there was a word that I would use to describe this collusion, it's a process of de-democratization. We spent hundreds of years getting away – at least in the West – from both monarchy controlling and organizing society, and the church organizing and controlling society, and we spent hundreds of years establishing bureaucracy to get us to this point where we have routines and processes to govern society, and we are at a point of inflection, this collusion between state and private power, and specifically technological power. And so, for me, the question I hear, from a movement perspective, is a question that Tamika Lewis, one of the co-founders of Our Data Bodies, has asked before, which is 'When the surveillance state is abolished, what's the first thing that you're going to build?'. I think it's a beautiful question, partly because we believe that we can abolish these systems of control. So it's an affirmation that refusal is possible; it might not be immediately possible within our lifetime, but there are certainly incremental things that you can do, some of which Patrick has referred to. But then, the question of what is the first thing you're going to build is really important to think about. It's not that the big tech is just going to go away. Or institutionalized racism is just going to go away. We need to build a new way of organizing society.

And that is a really complicated question to answer – I don't know if I have a good answer. But I have thought about how do we organize knowledge production in a different way. Maybe away from the 'university' because university has had a monopoly for way too long and it's gotten pretty inside-out at this stage – perverse incentives motivating higher education, for example, in the United Kingdom or the United States. So how we organize knowledge is the first thing that I would want to help build. And imagining that space and practising, rehearsing for that moment, I think, is part of what we are trying to do with Our Data Bodies and some of the other work that I'm engaged with. It's thinking about how you create alternative ways of thinking and doing and that can feel really abstract, unreachable, but I think it's possible.

When I think about my own role, if I get really specific, because I am the one situated within academia on the team, one specific thing that I've thought about in terms of digital refusal that maybe I could seed – is subversive practice within the computer science field. So I have thought, can we build a school for computer scientists ... who maybe are in secondary school or just starting university ... How do we build something that engages this epistemic community, and thinking about how to either track, monitor, get inside the guts of these big tech companies in order to expose what's happening; or undermine what's happening? How do we create a community, for example through a summer school, where computer scientists are ready and have the support system they need to whistleblow? And really try and get outside of this giant octopus, that we know now is really difficult to protest. If we look at the high-profile people who have protested what tech companies are doing, like Timnit Gebru or the myriad, usually unnoticed, Black and Brown techies that quit the tech industry and talked about severe problems that are happening – we should be able to coalesce those people, in a sort of underground school that gets them ready for long-term observation and interference in this beast that is de-democratizing society. That's one specific thing, I am ready to move that forward.

Patrick Williams: There is something quite unnerving about a room for the computer scientist plotting on how to unravel Amazon. I would not want to be in that room. But there is something about – and maybe again being quite cynical about this, and something that I find myself grappling with – is that sense that refusal assumes an understanding of the tech and the harms that accompany the tech. And I am not sure we are there yet. I am not sure there's enough there for us to even convince individuals of these arguments. There is a piece there around knowledge and community-based understanding and appreciation of this. And it is because at times I have also heard members of communities where I have worked, especially in response to, for example, questions of violence, interpersonal violence, advocate for the use of tech within communities.

I sat speaking with a community worker who works with young people, recently, who spoke about the need for more CCTV cameras within the neighbourhood and within the community, as there is a need to identify those individuals who are responsible for interpersonal violence. And as you can imagine that led into a lengthy conversation between us. But there is that sense of how do we insulate and protect kids if we are not using these tools? The seduction of technology as a way of improving the likelihood of detection of individuals who may pose harm. Don't get me wrong, I recognize, we all recognize state/structural harms which are bearing down and driving and facilitating harm at the micro level. But in essence it's when the community demands some of these tech solutions. And I guess that's where I am in a bit of a bind … where I begin to reflect on how do we begin to move this forward?

I remember having a conversation in the research team I was involved with around the question of what technologies do we abolish? Or should we begin to advocate for the abolition of tech, or certain forms of tech, and what do we accept? I remember in that conversation an individual saying that facial recognition is a [definite] no. That's a tech that we would argue should be abolished. It was really interesting hearing that, and these were techie folks beginning to think around this hierarchy of what remains and what goes. Again, there were implications around that very question. And it's forcing me to attempt to reflect on some of this. But I do think that the question you asked – once the surveillance state is gone, what is the first thing to be built – is a really powerful question. But then I'm also reminded by Ruthie Gilmore's line … that abolition isn't about was absent, but what is present. So a part of me is thinking about what do we need to build now, facilitating the abolition of the surveillance state – which is also the point you are making, Seeta.

Back to the refusal with regard to the use of tech – I am thinking about the group of computer scientists I worked with, who spoke about 'piggy leaks'. It was a build on the WikiLeaks thing, and was about how do we begin to use tech as a way of identifying those individual police officers who pose harms, or criminal justice systems or agencies that pose the most harm. And what data can we draw from, what fusion of data can we build as a way of building risk predictors of those areas, those officers, those institutions that pose the most harm for racialized communities? And again, we've gotten into additional conversations around well, is it about appropriating the tools and languages, as a way of trying to facilitate change, or do we have to break through, and away from those tools? Again, I feel like I'm posing more questions, rather than offering solutions, and I think that's appropriate in this moment because we are moving towards this intensification, this consolidation of tech.

We are seeing the legislation being built to support the exclusion of those that are deemed as risky, suspicions. So, I guess it is about thinking through what can be done, how can we begin to imagine the absence of a surveillance state in this moment? I guess, for me, this is a sense that refusal has to start from a sense of knowingness; it's about knowledge production, how do we begin to build a sense of a knowingness so that individuals become aware that to have the app and be able to go out to eat is giving over much more in that moment. And I am not sure to what extent we know this. I hear people who get it, but I'm not sure to what extent the vast majority of individuals are really aware of the potential harms that are coming along the road which we are literally walking into. So there is a question around this knowingness and how do we begin to build this knowingness as a way in which individuals can, through the notion of informed consent, begin to refuse the tech. Refusal from the place of knowing what's coming and this is why we need to refuse. Does that make sense?

Seeta Peña Gangadharan: It does really make sense to me! You folks sparked many different thoughts in my head. You said it is about knowing, and creating a system of knowing. And it reminded me of a project that Mariella Saba wanted to create for a long time in her neighbourhood. It is slightly different from the 'piggy leaks' idea, but is also thinking about alternative systems of knowing and being. So she had this idea to convert an old police station into a youth centre, a cultural centre for young people. There's something that's haunting about that, but also really beautiful about that concept. Haunting because of all of the spirits and memories that live in that place having dealt with the police. But something beautiful that comes out of that horror, that's about re-establishing a connection with one another. Because for too long we have been atomized in how we deal with the lack of security or safety in our communities, the lack of feeling like we belong in our communities.

So now we have three kinds of examples: the kind of subversive underground school [for tech specialists], 'piggy leaks', and this other idea of kind of converting, transforming an institution into something new that affords a different sense of safety and belonging. I guess what that tells me is that there's never going to be just one path. The answer is going to be in many different types of movements and actions and practices that are cultivated, that appear and disappear, or that may be staying for the long term. I think in some sense that's part of what's going to move things forward. And don't get me wrong, it can feel really down at moments, especially when I am reminded of the two settings that I encounter. If I have an opportunity to go to LA while I'm here in the USA, the problems that we're talking about such as over-policing and hypersurveillance are immediately present. And then when I'm sitting in a weird meeting with academics and policy makers who are making these seemingly abstract decisions, that have implications for the whole of communities, like in Los Angeles … It feels stuck, like nothing is going to change – you just feel the weight of someone saying 'Oh yes, let's just get some [surveillance] cameras. And just thinking how are we going to untangle this?' It can feel really grim. And at the same time, re-imagining if we were to create that youth centre that Mariella is talking about. If we are to move forward with piggy leaks and other kinds of ideas. That is exciting. That's the human imagination expanding, and it is really motivating, you start to see different possibilities.

Patrick Williams: I was just thinking about that idea of refusal. There is a really important notion of refusal, but also of what individuals need to know to be able to refuse. But I always get pulled back to that idea of an encounter between an individual and the police officer, or a probation officer. And what does refusal represent in that moment? And what can often result is an escalation, in terms of increasing the likelihood of force, or criminalization, or breach of custody. And so, I am wondering whether the extent of refusing the digital increases the likelihood of you being exposed to, or encountering, the tech? There was a pilot where they were testing the facial recognition tech, and if it didn't recognize an individual, then that prompted the police to stop the individual. So even notions of refusal, or the system refusing to acknowledge or see the individual, may result in the individual having to therefore encounter the police. So again, I am thinking about what does refusal represent, and I quite like the idea of just misspelling the name, or just taking out the digit of a mobile number. But there is still something about how deliberate refusal will often resolve in an escalation in that encounter, and that will almost certainly increase the likelihood of individuals' details being captured and held and then manipulated by the state.

Seeta Peña Gangadharan: Yes. The analogy for me is in the work that I was doing around supporting the communication needs of different social justice and social change groups, and part of that work was learning the grammar of mainstream media. How to include, how to exclude, how to exploit.

Learning what to anticipate. And I feel like we are definitely learning the grammar of police tech. Police tech in the traditional realm of policing, and then police tech, as you were describing, Patrick, in terms of this hyper-normalized everyday policing which is now about your vaccine passport to get into the restaurant to get some food. Or to travel. And so we are learning a grammar. We are learning that certain actions will cause a reaction. And we can anticipate. We know a little about human behaviour. We can establish patterns. Make it harder as things become more fully automated. But I am not the type of person to say that we are now dominated by robots. I don't think we're going to be in a fully automated society. We are always going to have this human element there. But we can anticipate that.

There was a question that once appeared in a workshop that I gave to activists and organizers around digital security, and somebody actually asked: 'If I download Signal, or if I use this other privacy protecting device, is that going to increase my chances of being surveilled?' And that is actually a great and valid question because you can be surveilled in your absences; your absences speak a lot about you, people can infer your behaviour; computers can compute on the basis of that. But we can also anticipate that. In computer science, and especially in privacy and security research, everybody's talking about this war of escalation of all these techniques that you do to outsmart the privacy-enhancing technologies that have been created, always about gaming the system. That is adversarialism at its core. And I think you can do that and, alongside of that, there's also 'Let's just do something different'. Because y'all are paying attention to this right now but I'm going to be over here, making something new and you're going to have to figure out my logic soon. I think people in communities start to figure that out by virtue of having to survive through that.

Patrick Williams: There is a sense of giving over, however, your authentic self in the data itself, in gaming the system.

I recognize that what you do is trying to keep away your authentic self from the data self that you may need to present to the police officers, or employers, and try to retain some sense of Patrick, here, as distinct and different from the data self. Just thinking about how we present ourselves online in the digital world. And how we resist, and the authentic self sits somewhere else.

Seeta Peña Gangadharan: I guess in hearing you speak I'm just struck by what we're coming to anticipate. If the state historically has a monopoly on violence, I think that is changing to this intersection between the state and technology companies. And learning collective self-defence is difficult but necessary. We are seeing a new way of monopolizing violence in society.

Patrick Williams: I would probably go beyond necessary, and say it's crucial.

Seeta Peña Gangadharan: Yes, urgent. And you would never know that if you just read the PR strategy of Amazon or Microsoft. Microsoft is the best example because they have got off the hook so many times over, yet their power, a lot of this stuff on facial recognition tools;, they have a cloud infrastructure. But you would never know. Violence isn't a word that's part of their PR strategy or owning up to any dimension of violence in the work that they do. This tells me that we need a new discourse around what technology is.

Patrick Williams: I suppose this is precisely why the language of 'data harms' needs to be extended. It is not simply about how the tech may impact on an individual basis, you see structural effects. And I think the language of violence is appropriate here, and beginning to think about it that way drives the urgency around this notion of self-defence and the need for self-defence. But also opens up other avenues … it forces us to reimagine, what does this self-defence begin to look like and feel like?

As the transcript hopefully demonstrates, what started off as an 'interview' very quickly became a fiery and incredibly powerful conversation that centred around the biopolitics of digital pasts and futurities. We hope that some of the themes – which are really the tips of monumental icebergs that need urgent attention – will serve as an inspiration to researchers and activists alike, and open further dialogues surrounding questions of pandemic and post-pandemic digitalities, (in)equalities and social justice.

References

Benjamin, R. (ed) (2019) *Captivating Technology: Race, Carceral Technoscience, and Liberatory Imagination in Everyday Life*, Durham, NC: Duke University Press.

Williams, P. (2018) 'Being matrixed: the (over)policing of gang suspects in London', *Stop Watch: Research and Action for Fair and Inclusive Policing*. Available from: https://www.stop-watch.org/our-work/gangs-matrix [Accessed 18 May 2023].

Epilogue: Digital Disengagement – Questions of Pandemic and Post-Pandemic Digitalities

Adi Kuntsman, Sam Martin and Esperanza Miyake

This book began with a critical question – 'What is the place of digital refusal in the fabric of pandemic and post-pandemic life?' – which emerged as part of our collective reflection on the rise in digitization during the first two years of the COVID-19 pandemic, and beyond. The process of answering this question was often accompanied by the rapidly diminishing possibility of stepping away from compulsory digitalities. Presenting the paradigm of digital disengagement that challenges the normalization of the digital (and of digital solutionism in particular) and foregrounds opting out and refusal as a starting point, the book thus explores a number of aspects of digital disengagement in pandemic and post-pandemic times. Some of our contributors have focused explicitly on the social and political effects of forced digital engagement; for example, Esperanza Miyake's discussion of its macro-level global economic and racializing power, or Serra Sezgin's analysis of its impact on micro-level working practices and work–life balance. Others have documented and analysed practices of disengagement and refusal, and the paradoxes they involve. Sam Martin's chapter explored digital disengagement, which relies on furthering one's digital skills and data literacy to avoid datafication, state surveillance and potential criminalization through use of fertility apps, whereas Chelsea Butkowski detailed how over-saturation with digital connectivity during the pandemic led many to step away from the digital almost entirely into the domain of pen and paper – only to share the results online afterwards. Continuing the topic of digital disengagement as a means of resistance to surveillance, Annika Richterich turned to hacker practices of data minimalism as a method to systematically, consistently and deliberately reduce the level of invasive datafication. Here, digital disengagement is less about a break away from the digital, but instead is about its systemic reduction. Moving to exploration of digital disengagement in the context of climate crisis and ecological harms, Adi Kuntsman's chapter looks at systemic forms of digital reduction to reduce the toll of pandemic digitization on the environment. Concluding and expanding the discussion of digital disengagement in contexts of systemic harms, Seeta Peña Gangadharan and Patrick Williams discuss the possibilities of digital disengagement as resistance to state power and the violence it inflicts on racialized communities.

Although this is a short collection that looks at a relatively select number of international contexts, the discussion of digital disengagement it sets up has broader and long-lasting implications. For instance, one of the urgent themes for further exploration that emerges from this book is the need to discuss digital disengagement globally, and through a combination of economic and political frameworks. There is a growing need to understand the inter-relatedness of the economic power, political formations and global inequalities that are involved in maintaining the (post)pandemic digital dependency at all levels from countries to communities to individuals. One particularly telling example of how much the global economy, connectivity and social mobility has become dangerously dependent on digital engagement – to a point of threatening life when digitality is removed – is the case of China's zero-COVID policy, adopted throughout the pandemic. While localized in response, because of large tech companies like Apple that have factories located in China, the country's first lockdown caused a disruption in the global supply chain (Farrer, 2022; Song, 2022). This example sheds a sharp light on the unequal distribution of 'global' digitality as defined by the global assembly line (Qiu, 2016): one where there is a marked difference between the consuming Global North/'West', where this crisis is in the lack of consumer products flowing into Western countries in time for Christmas; and the productive Global South/'Other', where the same crisis is about social welfare and the infringement of human rights (for example, when factory workers are forced to sleep in factories as part of quarantine measures). Digital disengagement, as this example demonstrates, and as we show throughout the book, is defined through the geopolitics of the economy, which makes digital refusal an action that is at once subject to both localized and globalized forms of disciplinary control over citizens, users, workers and consumers/producers.

At the time of writing, it has now been three years since the COVID-19 global pandemic began. Yet our discussion of how we can resist pandemic and eventual post-pandemic digitalities – on multiple local and global levels – is more relevant than ever. One of the key theoretical contributions that this collection makes is in its critical assessment of digital solutionism, its pitfalls, its dangers and its failures. And while three years is only a short period, it nevertheless allows us to both see the rapid changes that took place, and to ask what's next. For example, over the course of the pandemic, we saw at first a reliance by Western governments on digital solutionism or a 'digital-first solutionist approach' (Rowe et al, 2020) to keep track of COVID infections, and indeed vaccinations, in the form of contact-tracing apps, check-ins and then vaccine passports (Marelli et al, 2022). However, pushback by citizens grew with regard to the efficacy of the digital contact-tracing apps, for example, and potential invasion of personal freedoms in the form of imposed regulations and mandates (Martin and Vanderslott, 2022). At the same time, governmental needs to save on resources used to implement digital public health measures in the early days of the pandemic saw a gradual turn away from digital solutionism in an attempt to 'live with COVID' (Limb, 2022). Interestingly, many digital practices – from contactless payments to digitization of public services, registers and travel – which were initially introduced as pandemic safety measures, have remained, serving state and financial interests. Yet, in many Western governments' approach to public health, digital measures gave way to more traditional methods such as public information campaigns for vaccine uptake, and general 'common sense' advice regarding masks or self-isolation upon getting COVID (Stokoe et al, 2022), despite the considerable financial investment that had gone into developing digital solutions.

One may argue that this can be seen as a 'non-individualistic digital disengagement', occurring at the macro level of government and public health systems (rather than the level of personal or institutional choices), in response to citizen criticism and resistance. Yet, this move is much better understood as a form of state abandonment and disinvestment, masquerading as digital reduction for reasons of efficiency, or even as a reaction to public concerns about privacy. In many countries, such digital disinvestment co-existed with other forms of state withdrawal from financing COVID-19-related public health measures, such as access to free lateral-flow and PCR tests, extended sick leave, or financial support for individuals and businesses affected. Such a move made COVID safety a matter of resources and wealth, or the lack thereof. The directive to 'live with COVID' is particularly affecting those with compromised health and the disabled, as well as the poor, the financially disadvantaged and the precarious.

The distinction between macro-level digital disengagement as a remedy for digital harms and as a form of state abandonment and withdrawal of responsibility is particularly important. This is because we often see that compulsory digitality is introduced – and institutionally enforced – side by side with a withdrawal of digital services or centrally restricted access to information, via online censorship and internet shutdowns. While politically motivated forced disconnection has been widely discussed (Vargas-Leon, 2016; De Gregorio and Stremlau, 2020; Volpicelli, 2021), our book brings to the fore another potentially crucial reason for withdrawing digitality: the finite nature of resources (such as rare metals or electricity) that are required to produced digital communication devices and sustain the communication itself (Qiu, 2016; Cubitt, 2017; Maxwell and Miller, 2020; Greenwood, 2021). Between the climate crisis and various ongoing economic crises, especially around oil, gas and energy prices, digital withdrawal might become a reality in the near future, whereby the resources needed for global digital connectivity may be barely affordable and possibly not even available at all.

We therefore conclude this book with an invitation to critically re-evaluate the power, promise and affordability of compulsory digitality as we move towards a post-pandemic future, and consider multiple possibilities of digital disengagement – ones that are driven by sustainability, equality and justice, rather than economic gain, political control or withdrawal of care. We hope that this book has paved a way to such considerations.

References

Cubitt, S. (2017) *Finite Media: Environmental Implications of Digital Technologies*, Durham, NC: Duke University Press.

De Gregorio, G. and Stremlau, N. (2020) 'Internet shutdowns and the limits of law', *International Journal of Communication*, 14: 4224–4243.

Farrer, M. (2022) 'Zero-Covid policy is costing China its role as the world's workshop, *The Guardian* [online] 3 December. Available from: https://www.theguardian.com/world/2022/dec/03/zero-covid-policy-is-costing-china-its-role-as-the-worlds-workshop [Accessed 5 December 2022].

Greenwood, T. (2021) *Sustainable Web Design*, New York: A Book Apart.

Limb, M. (2022) 'Covid-19: Is the government dismantling pandemic systems too hastily?', *BMJ* 376: o515.

Marelli, L., Kieslich, K. and Geiger, S. (2022), COVID-19 and techno-solutionism: responsibilization without contextualization?' *Critical Public Health*, 32(1): 1–4.

Martin, S. and Vanderslott, S. (2022) '"Any idea how fast 'It's just a mask!' can turn into 'It's just a vaccine!'"": from mask mandates to vaccine mandates during the COVID-19 pandemic', *Vaccine*, 40(51): 7488–7499.

Maxwell, R. and Miller, T. (2020) *How Green Is Your Smartphone?*, Cambridge, UK: Polity Press.

Qiu, J.L. (2016) *Goodbye iSlave: A Manifesto for Digital Abolition*, Champaign, IL: University of Illinois Press.

Rowe, F., Ngwenyama, O. and Richet, J.L. (2020) 'Contact-tracing apps and alienation in the age of COVID-19', *European Journal of Information Systems*, 29(5): 545–562.

Song, W. (2022) 'What is China's zero Covid policy and what are its rules?' BBC News [online] 5 December. Available from: https://www.bbc.co.uk/news/59882774 [Accessed 5 December 2022].

Stokoe, E., Simons, S., Drury, J., Michie, S., Parker, M. and Phoenix, A., et al (2022) 'What can we learn from the language of "living with covid"'?, *BMJ*, 376: o575.

Vargas-Leon, P. (2016) 'Tracking internet shutdown practices: democracies and hybrid regimes', in F. Musiani, D.L. Cogburn, L. DeNardis and N.S. Levinson (eds) *The Turn to Infrastructure in Internet Governance. Information Technology and Global Governance*, New York: Palgrave Macmillan.

Volpicelli, G.M. (2021) 'The draconian rise of internet shutdowns', Wired [online] 2 September. Available from: https://www.wired.co.uk/article/internet-shutdowns [Accessed 18 December 2022].

Index

References in **bold** type refer to tables.
References to endnotes show both the page number
and the note number (231n3).